a guide to hi-fi

a guide to hi-fi

John Wasley and Ron Hill

PELHAM BOOKS

First published in Great Britain by
PELHAM BOOKS LTD.
52 Bedford Square
London WC1B 3EF
1977

ISBN 0 7207 0906 7

Printed in Great Britain by A. Wheaton & Co., Exeter

contents

1. Sound and System 11
2. The Amplifier—Heart of the System 27
3. Tuning in to Tuners 43
4. Gramophones—Up to Date 57
5. Adding Tape 89
6. Where the Sound Comes Out 115
7. Be a Hi-Fi Doctor 131
8. Caring for Long Life 157
9. Vocabulary for Hi-Fi Addicts 171
Index 185

Acknowledgements

The authors would like to express their gratitude to the many people who helped, advised and guided us in the production of this book. They are too numerous to be listed and, in any event, would be slightly embarrassed to see their names in print. However, particular thanks are extended to Jean Wasley, who so carefully read through the manuscript, correcting errors and commenting on certain aspects; to Jean-Marie for lending us her eraser; and to the following companies, listed alphabetically, who generously provided us with illustrations.

Alba (Radio & Television) Ltd., Bull Lane, Edmonton, London N 18.

Armstrong Audio Ltd., Warlters Rd., London N7 0RZ.

Basf United Kingdom Ltd., PO Box 473, Knightsbridge House, 197 Knightsbridge, London SW7 18A.

Bib Hi-Fi Accessories Ltd., PO Box 78, Hemel Hempstead, Herts. HP2 4RH.

CBS Records Ltd., 17-19 Soho Sq., London W1V 6HE.

Heath (Glos) Ltd., Gloucester.

Natural Sound Systems Ltd., Strathcona Road., Nth. Wembley, Middx.

Photax Ltd., Eastbourne, Sussex.

Rank Audio Visual Ltd., PO Box 70., Great West Road, Brentford, Middx.

Research Institute Ltd., Kernick Rd., Penryn, Cornwall.

Rola Celestion Ltd., Ipswich, Suffolk.

Shure Electronics Ltd., Eccleston Rd., Maidstone, ME15 6AU.

SME Ltd., Steyning, Sussex.

Strathearn Audio Ltd., 109 Jermyn Str., London W1.

Cecil E. Watts Ltd., Darby House, Sunbury-on-Thames, Middx.

List of Photographs

(between pages 96 and 97)

Photax-Concertone 800B stereo amplifier.
Alba UA 900 stereo amplifier.
Armstrong 623 AM/FM tuner.
Country singer, Wayne Nutt, operates a mixing console at CBS recording studios.
'Stampers' ready to go into action at CBS record manufacturing plant.
Strathearn SMA 2 electric servo direct-drive turntable.
SME pick-up arm.
Shure M95ED cartridge.
Akai 1722L stereo open-reel recorder.
Yamaha TC 800GL stereo cassette recorder.
Basf 8200 stereo cassette deck.
Nakamichi TT 700 three-head cassette deck.
Heathkit TM-1626 stereo microphone mixer.
Celestion Ditton speakers showing cone dispositions.
Yamaha HP1 headphones.
Groovac vacuum record cleaner.

List of Line Illustrations

		page
1	The middle and inner ears.	13
2	The wavelength.	14
3	Compression and rarefaction.	16
4	Frequencies compared.	17
5	Frequencies (including overtones) of musical instruments compared.	19
6	Spectral light behaviour.	19
7	Amplitude.	20
8	Cartridge and cassette.	24
9	Signal behaviour in an amplifier.	31
10	Simple circuit	34
11	Plotting the root mean square of a signal voltage.	35
12	Frequency response curve.	37
13	Amplitude modulation.	44
14	Frequency modulation.	45
15	Tuner dial.	49
16	Chart of radio signal progress in a superheterodyne receiver.	55
17	The record-making process.	59
18	Mixing musical sounds.	61
19	Recording and equalisation curves.	64
20	Electronic servo speed control.	71
21	Bias, the natural movement of a pick-up arm.	77
22	The fulcrum principle.	78
23	The geography of a pick-up arm.	78
24	Close-up of a stylus in a record groove.	82
25	Stylii—plan view.	83
26	Bib stylus balance.	84
27	Hafler effect, with one speaker and two.	86

28 Quadraphonic encoding and decoding. 87
29 Open-reel tape recorder head positions. 94
30 Stereo tape tracks. 98
31 Stereo cassette tracks. 99
32 Cassette recorder head positions. 102
33 The Dolby system at work. 104
34 Moving coil, or dynamic microphone. 105
35 Tape splicer. 110
36 Speaker positioning for stereo listening. 124
37 Three-pin plug showing wiring. 133
38 Circuit diagram of the Heathkit AM-FM solid state stereo receiver, model Ar-1500A (power supply circuit only). 135
39 A resistor with colour coding. 137
40 A transistor. 138
41 Two types of capacitors. 139
42 A diode. 140
43 The multimeter. 143
44 The banana plug. 151
45 The DIN plug. 153
46 The phono plug. 154
47 Bib record 'Dust-Off'. 161
48 Bib Groov-Kleen record cleaner. 161
49 Tape cleaning outfit. 164
50 Cassette head cleaner. 165
51 Cartridge head cleaner. 166
52 Bib cassette storage cabinet. 167
53 Bib cassette fast hand winder. 169

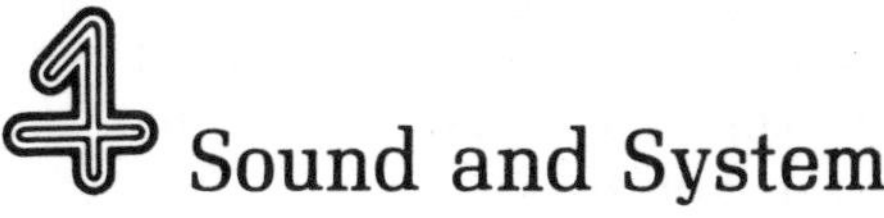

4 Sound and System

So you want to know about Hi-Fi. That is understandable since it is so popular and is not an easy subject to comprehend. No doubt you have discovered that there is a little more to obtaining high fidelity sound than simply calling on your local dealer and picking up a nice-looking outfit.

Appearance is all very well, but that is probably the least important part of the problems of Hi-Fi. What you are really after is a good sound.

The trouble is, how do you define a good sound? What is abysmal to one person is perfectly adequate to another. Other difficulties impose themselves on you too. For instance, when you listen to Hi-Fi outfits in the shop, they all make beautiful, silky music in your ear. But, when you get your carefully selected turntable, amplifier and speakers home, you cannot quite produce the same sound.

What went wrong? Is it the acoustics? Perhaps it is the position of the speakers. The truth of the matter is, any number of factors could influence the quality of the sound which reaches your ear.

But, let us step back even further. Perhaps you are about to buy an outfit. Because you have not spent half a life-time working with electronic equipment, you are having trouble threading your way through the plethora of jargon and the maze of Hi-Fi outfits.

Well, you are in luck. You have come to the right place. Within these pages you will learn all about the real meanings behind the electronics experts' mumbo jumbo. You will then be able to choose your Hi-Fi outfit with confidence—an outfit which will be tailor-made for you, and will make sounds which please you.

If, on the other hand, you have an outfit already and you want to know more about it, you will find what you want here too. You will learn what goes on inside your equipment. At least, you will know enough to give you sufficient knowledge to be your own Hi-Fi doctor. Diagnosing faults will be a simple job.

But, enough of that. You know something of how this book is going to help you, otherwise you would not be reading it. The matter of the moment is to get into the subject, and a good place to start is to embark on a voyage of discovery to lay bare the 'nature of the beast'.

THE HUMAN AUDIO SYSTEM

We are so used to hearing noise that we take it for granted. It is like electricity. The average person knows of its existence, uses it frequently, yet does not really care how it works. After all, it exists, it is useful, and there is nothing we can do about it.

Alright then, sound is there for all whose ears are in working condition to hear. Someone bangs a table, or turns on a radio and the sounds reach the human audio system. They are collected by the ear and channelled to the drum. This vibrates and transmits the reverberations through the malleus, the incus and the stirrup (the small bones of the ear) to the cochlea (the inner ear). From there the auditory nerves carry the sound to the brain.

The question is, how does the sound reach the ear in the first place?

Sound waves revealed. Air, is the answer. Sound needs a medium through which to travel. As it originates in the form of a vibratory motion, it transmits itself through the air in this way.

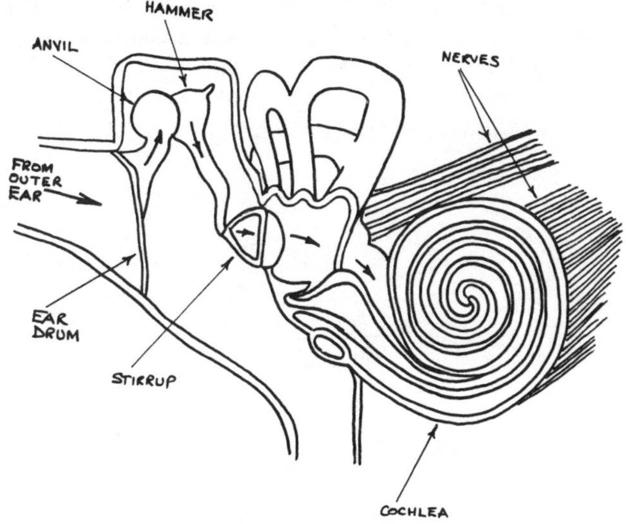

Fig. 1 The middle and inner ears.

It is much like dropping a stone into a flat, calm pond. The impact of the stone hitting the water creates a series of circular ripples which appear to move outwards from the point of impact. The further these ripples are formed from the centre, the weaker they are.

In point of fact, the water itself does not move in a lateral motion. It merely moves up and down as the 'vibration' travels outward. It is important to understand this fact because sound waves behave in exactly the same way in the air.

The vibration moves outwards from the centre of the sound, moving air currents. This sound weakens too as it moves further away from the source.

You have probably seen the drawing in the *Radio Times* of the broadcasting mast surrounded by a series of circles. These represent sound waves, but look very much like the stone in the pond example, demonstrating how alike the two actions really are.

Vibrations by the number. Now, it is possible to vary the number of sound 'ripples' being emitted so that they are closer together or more distant from each other. This is where we come to the first technical term. The inter-wave distance is known as the wavelength.

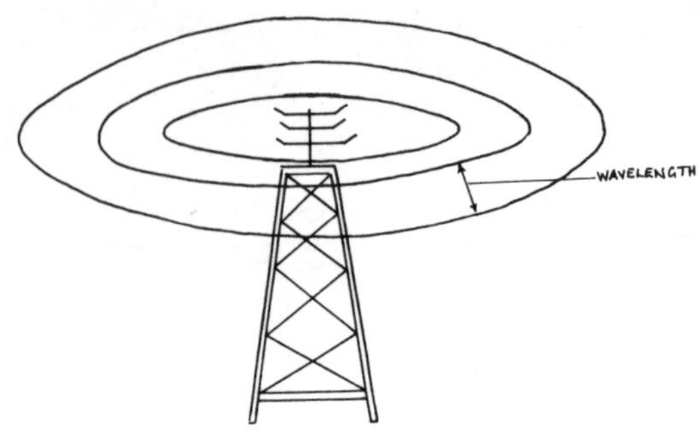

Fig. 2 The wavelength.

Inasmuch as these sound vibrations travel, the wavelength has a bearing on how often the peaks of these vibrations are formed at a specific point within a specific period (it moves like a series of ripples on a pond, remember). If the wavelength is short, then the number of 'wave' peaks forming at your specified point will be more frequent than if the wavelength is long.

By determining the number of peaks formed at your chosen point within the specified period (usually one second), you are able to find the frequency, or pitch. When the peaks are closer together, the frequency or pitch is said to be high. When they are well separated, they are known to be of a low frequency or pitch.

Try an experiment. You can see this work for yourself. Take a well-sharpened pencil and tie it in a vertical position to the lowest sounding string on a guitar. The point should be downwards and touching the body of the guitar. Slip a piece of paper under the point, pluck the string once, and draw the paper steadily and evenly along the length of the string.

You will see a frequency pattern appearing before your eyes. It is fun, isn't it?

Now draw a straight line through the centre of this frequency pattern and you will see that the waves are even on either side.

The next step is to repeat the experiment with the highest-pitched string and see the difference. While both will be even both sides of the line, see how the 'waves' are much farther apart in the first experiment, than in the second. The frequency pattern resulting from your second experiment then demonstrates that the highest sounding string is a high frequency or pitch.

Compressions and rarefactions. So far we have been looking at these sound waves from a plan view, so to speak. That is from above—or below, depending on how tall you are. Now let us have a look at the side view.

From this direction too, the sound 'ripples' look like waves. There are peaks and troughs which form progressively further away from the disturbance as the sound vibrations move outwards.

Name-droppers, stand by. The peaks of the 'waves' cause increased pressure in the air and are called compressions. The troughs cause a decrease in air pressure and are termed rarefactions.

Sounds compared. As we have already learned, all sounds are not the same. A violin sounds nothing like a double bass, for instance. Also, there is no mistaking the difference between a gun shot and a baby's cry. This may seem obvious, but it is important to know that the resultant sound waves are different too.

For instance, if you were to take the double bass and the violin (or indeed, the high and low strings of a guitar), and compare the sound waves, they would be distinct from each other. You would be able to guess which was which merely by looking at a diagram of the sound waves. The violin's pattern would show the peaks very much closer to each other than the double bass frequency pattern.

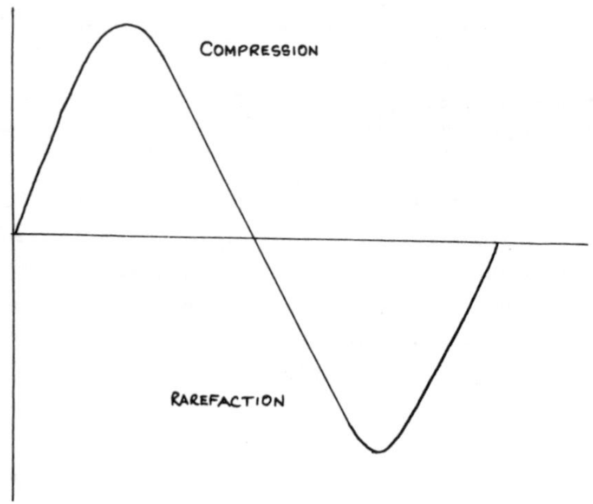

Fig. 3 Compression and rarefaction.

Measuring frequencies. To leave it like that would make the study of frequency patterns a hit or miss occupation. Precise measurement enables the experts to determine exact frequencies—so vital for all aspects of electronics.

In point of fact, the word 'frequency' denotes a term of measurement. The standard of measurement is generally over a time scale of one second. You simply study the sound waves and note the number of peaks which pass in that second and the answer is the frequency. However, it is also known as cycles per second—c/s for short.

You read just now about the difference in noises—the violin and the double bass, the gun shot and the baby's cry, and so on. Well, the gun shot may make a single report, but a baby certainly does not make the same noise on the same frequency or pitch all the time. It ranges from a low moan to a screech.

The human voice can range over a number of frequencies. This is how we are able to make inflections in our sentences, and to make our feelings felt when we are upset. On average, a human voice can spread itself from 40 to 10,000 cycles—that is, from forty peaks a second to ten thousand!

But, that is not all. Our ears are much more versatile than our voices. They can cope with anything from 30 c/s to 18,000 c/s in most cases. However, our hearing ability is dependent upon age. The older we get, the less we can hear. The highest frequencies our ears can handle may drop to something like 3,500 c/s at 50 years of age.

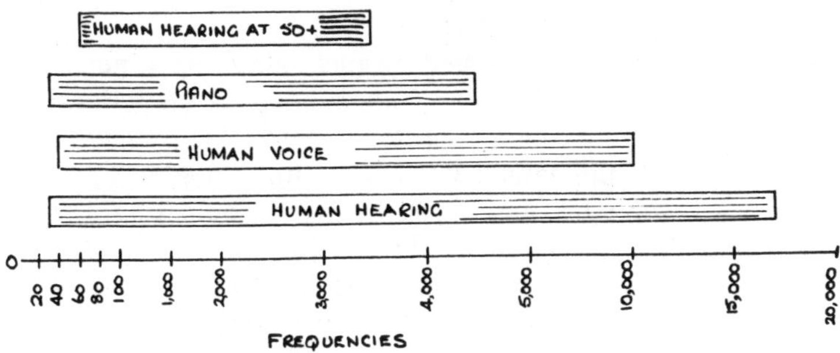

Fig. 4 Frequencies compared.

Musical instruments have a wide range too. They make music by playing a variety of notes on the frequency scale. Of course, it has to be precise, and it is necessary to select specific frequencies or notes in order to make a tune which is pleasant to the ear. A piano, for example, is able to select notes in a scale which ranges from 30 c/s to 4,200 c/s (approximately).

In addition, it is possible for sounds to be emitted from devices which are pitched too high for the human ear to receive audibly. Such a device is a dog whistle.

When a frequency is expressed in thousands of cycles per second, it has a special term. It is known as 'kilocycles'. When it reaches the millions of cycles per second (not at all uncommon), the term 'megacycles' is used. Shortened, they appear as Kc/s and Mc/s respectively.

If you can remember the good old steam radio, these terms will be very familiar to you. But it has all changed now. Although we are currently running on two systems and Kc/s and Mc/s still appear, the continental term 'hertz' is fast becoming standard in the vocabulary of the electronics expert. It too has a short version—Hz. However, the prefixes kilo- and mega- still apply with the same meanings.

Nevertheless, despite the change of word, the standard of measurement has not altered. One hertz is identical to one cycle per second.

THE COLOUR OF SOUND

It is possible, as we have said, to pick out specific notes on any instrument. In fact they have been given letters to make them identifiable. Therefore, if you knew how, you could pick out the same 'A' on a violin, a double bass, a piano, a saxophone, and if you like, you could sing it too! As it is unlikely that you will have all these instruments in the house, you will have to set your imagination to work.

Think about the sounds emitted from each.

Even though it is the same note, they sound different, don't they? An 'A' on the piano, even with the best will in the world, will not sound anything like the same 'A' on the saxophone—or the violin, or the double bass, or your voice.

This is because each instrument has distinctive tones which surrounds its note(s). They are called 'overtones'. The overtones made by any given instrument are what makes that instrument distinctive, and are what gives its notes tone colour, or 'timbre'.

Perhaps the light spectrum is more familiar to you and can be used as a demonstration. White light is made up of a number of colours. These can be seen when light is passed through a prism. Water functioning as a prism produces the same effect and, as light passes through droplets in the air, a rainbow forms.

Now then, daylight is a great deal different from the light which comes from the 100 watt bulb in your living room. Daylight is bluer, yet both are still forms of white light. However, the colours which make up daylight are predominantly from the blue end of the spectrum, while household lamp light is an amalgam of colours primarily from the red end.

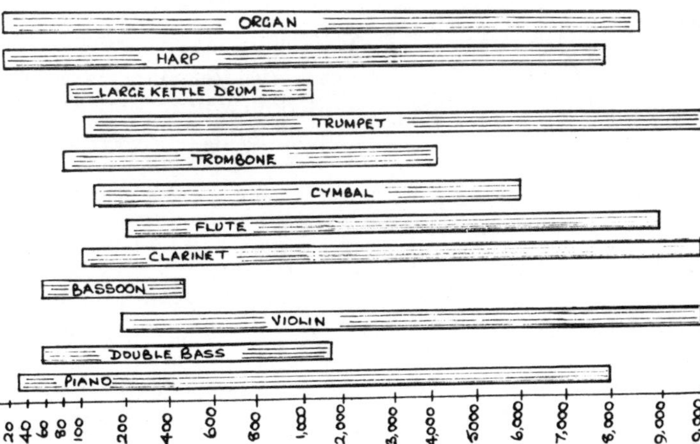

Fig. 5 Frequencies (including overtones) of musical instruments compared.

Because each type of light draws its component parts from different areas of the same spectrum, it becomes distinctive. Likewise, because each musical instrument draws its component parts surrounding any given note from different areas of the same frequency 'spectrum', it too becomes distinctive.

INTENSITY AND LOUDNESS

There is another dimension to sound which ought to be explored at this juncture, and that is the strength of the vibration. There is a word for that too, it is called

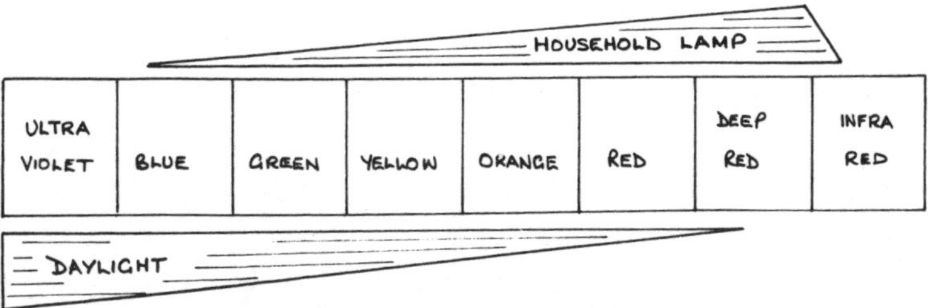

Fig. 6 Spectral light behaviour.

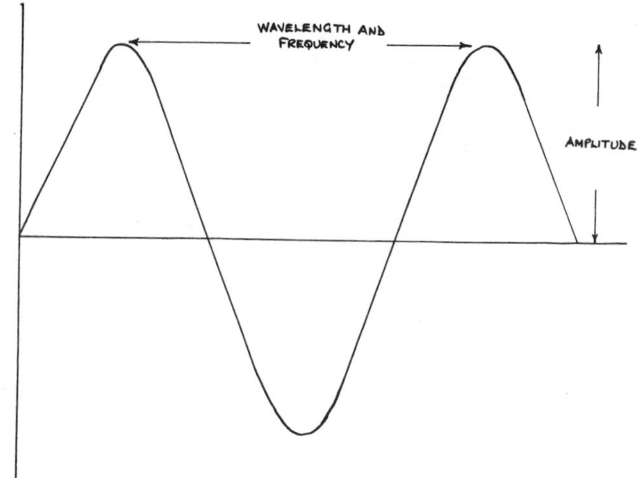

Fig. 7 Amplitude.

'amplitude' and is measured by the amount the 'waves' swing either side of the average point, or line.

Oddly enough, there is no direct link between the strength of the vibration (also referred to as the sound intensity) and how we hear it. The ear seems to have some sort of volume control which lowers the sensitivity (within limits) as the sound reaches it.

Therefore, the strength of a sound wave passes through two apparent stages. There is the actual 'intensity' which is the energy radiated, or power, of the unit emitting the sound, and there is 'loudness'—that is, how we hear it. Loudness takes into consideration the surroundings and the frequencies of the sound. As we hear some frequencies louder than others, the latter consideration is very important.

Ratios have been worked out to enable us to relate intensity and loudness. If the loudness is to be doubled, then the intensity must be multiplied by ten. These changes in intensity are expressed by the unit of the bel. A ten times multiplication or division of intensity equals 1 bel.

However, the bel is too large a unit to deal with, so it has been literally decimated. Units used to express the intensity of one sound against another are now measured in 1/10 bel, or the decibel (dB). As it is a logarithmic ratio, a sound which is double the strength of another is 3 dB stronger.

While dB are able to show a change in intensity, it is not suitable to describe a change in loudness. Remember that one of the criteria for loudness is the frequency. Therefore, an alteration in the level of loudness is expressed in the unit of the phon. A scale has been developed, at a point on the frequency scale of 1,000 cycles (1 kHz), starting from zero phon, which is the threshold of hearing, and ranging upwards. The point on the scale at which loudness causes pain is measured at some 130 phons.

THE REAL MEANING OF HI-FI

Having experienced the 'body' of sound all our lives we are now aware of the bones which give it form. Now let us have a look at what makes good sound.

The point is, what is Hi-Fi? To begin with, it is a contraction of the words 'High Fidelity'. As fidelity means truth, it is not out of order to construe that Hi-Fi means high truth—or, truthful sound.

In effect, it has been coined to indicate some sort of standard in reproduced sound. An excellent description is that High Fidelity sound should create for the listener the illusion that he is listening to the real thing.

Practically, this is not the easiest thing to do. Let us start with the record. The high quality sounds on a modern record are put there with very exacting and expensive equipment. It is designed to reproduce sounds as near to real life as possible.

But, there is no way of reproducing those sounds to create the illusion of the real thing without comparable equipment.

You can get an excellent example of the difference in reproduced sound quality by listening to a record on the best equipment stocked by your local Hi-Fi dealer, and then on your daughter's portable record player. Thirty seconds on each is all you will need. You will hear the difference, even with an untrained ear.

While this is an extreme example, it is representative of one of the major difficulties you will encounter when searching for a suitable outfit. Much equipment which is labelled 'Hi-Fi' cannot lay claim to anything like high fidelity reproduction. In an effort to present attractive outfits at attractive prices, the quality of sound suffers. In fact, the term 'Hi-Fi' often encompasses so-called 'Mid-Fi' and 'Low-Fi' equipment.

However, this is only usually the case for low-priced equipment. So if you are buying be careful of the glossy-looking outfit 'going for a song'. The rule when buying is, always listen and compare.

PUTTING AN OUTFIT TOGETHER

A complete Hi-Fi outfit is made up of several units. Any two or more could be in the same housing, but most enthusiasts prefer to have them separate.

It is, of course, much easier to purchase your outfit all in one, such as a radiogram or a 'music centre'. But to do so imposes certain restrictions on you. For instance, if the speakers are built into this all-in-one outfit you are considering, then you cannot get true stereo effects.

You will not be able to position the speakers where you want them to be, let alone at the correct distances for stereo listening.

Another major problem involves choice. When you buy an all-in-one outfit, you have to take what the radiogram or music centre has. While some of this may suit you—say the turntable and the tuner (radio)—you may not be happy with the tape recorder, nor the speakers, nor the pick-up arm. Unfortunately, there is little you can do about that. You have to accept all the units which come with the radiogram or music centre. You have no freedom of choice.

Finally, you have to consider the fact that no all-in-one radiogram or music centre has achieved the same sound capabilities and qualities as a carefully selected and matched outfit of separates. There are too many mechanical and electronic restrictions, you see. Only separate units allow you the choice to enable you to find that magic sound called Hi-Fi, and give you the illusion of listening to the real thing.

Having said all that about radiograms and music centres, it ought to be made clear, that they make very good sounds. But musical sound being such a highly personal thing, they cannot produce for you, your own personalised high fidelity.

Right, having decided to buy separates, you now have to look around. The first point to remember is that it is not necessary to buy all the same brand units. Provided you are able to match input and outputs, you can buy the unit which suits you the best. The majority of enthusiasts select one unit from this 'stable', and another from that 'stable', merely because each does the best job for them.

THE AMPLIFIER

Anyway, the first thing you will have to consider is the amplifier. This, as its name suggests, takes the tiny signal transmitted from the other units, and boosts it to an audible level. Amplifiers are available either on their own, or with a tuner built in. The choice as to which you buy is yours. There are no real arguments for or against. It is just a matter of preference.

THE TUNER

Should you want a separate tuner, they are available too. Simply remember that you are buying a radio. Find the best you can afford which will match your outfit electronically, and which will provide the range of frequencies (programmes) you are likely to require.

THE TURNTABLE

When you choose your amplifier, you will probably want to consider a turntable. This is the old familiar record player and is common enough not to require too much attention or explanation at this stage. As with every other unit, we will be exploring it in depth in later chapters.

In the meantime, when you try out turntables at your local dealers, play one of your own records and listen carefully. It will provide you with a standard against which you can measure sound quality on different turntables.

THE TAPE DECK

This is where making your decision can get complicated. Apart from having to select the tape deck for sound reproduction qualities, you will have to choose which tape system you want too. It really depends on your preference —cassette, cartridge, or open reel.

If you want to produce long, long programmes of music, then you ought to consider open reel. In addition to that, with the right equipment, open reel will reproduce the best quality sounds. However, for convenience, speed, and to allow you to use your taped music in the car, you will have to make your choice between either cassette or cartridge.

One point of interest, there are tape decks around which have provision for both open reel and cassettes. If you can find a use for both systems, it would be worth your while to look at, and listen to, decks such as these.

Fig. 8 Cartridge (left) and cassette (right).

THE SPEAKERS

Even with an amplifier, you will not hear any sound signals from any of the other units unless you have speakers. You will need two so that you can benefit from the stereo signals. If you are buying the new Quadraphonic system—that is, all-round sound—then you will need four speakers. Whatever the number, they should all be matched to each other and to the remainder of the Hi-Fi units.

Remember, the speakers are the units where the sound comes out. The resultant quality is very dependent on how the speakers perform. Make your choice carefully. Listen to several different types, preferably on the same equipment and using the same record. If you can get a home demonstration, so much the better. It cannot be emphasised enough that the speakers will make or break your Hi-Fi sounds—which makes them probably the most important link in the chain.

THE INDISPENSABLE UNIT

Do remember that these separate units are only functional when linked to the amplifier. You cannot, for instance, take the tuner only into another room. You will have to carry the amplifier and the speakers too—unless, of course, you buy another radio with an amplifier built in!

BUILD UP SLOWLY

All of this may seem like a lot to buy all in one go. But, do not be put off by the list. You do not need it all at once. You can gather your outfit together piece by piece.

For example, if you have a working radiogram already, it may be possible to use it as an amplifier and work your new record player or tape deck through it. If your radiogram is not stereo, then you will not get a stereo signal through it. But, at least you can start gathering your outfit together and use it right away. Later you can add new speakers and an amplifier—the rest can come later.

As you select your units be aware that you get what you pay for. The more money you have available to spend on your outfit, the better quality you can expect. Of course, there are variations within any price range and the trick is to get not just the best you can afford, but the best quality you can afford.

So, do not rush out and buy your outfit right away. Have a little more patience. Start looking around by all means. However, before you lay cash on your dealer's counter, read the next few chapters. They will help you gain a clearer understanding of how to choose the best outfit for your needs; how to get the best value for money; and, more important, how to get the best out of what you have.

See you on the next page.

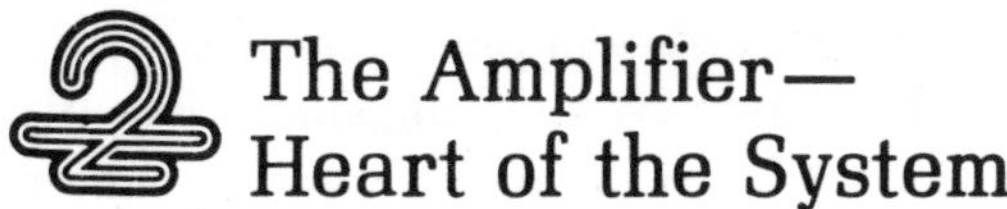

The Amplifier—Heart of the System

The heart of any Hi-Fi set-up is the amplifier. In fact, it functions very much like its organic counterpart. Voltage flows into the amplifier and is immediately 'pumped' out with sufficient force to drive the speakers.

If it was as simple as that, however, this would be a very short chapter. There is more, of course.

This sound-boosting facility requires that a fair amount of electronic wizardry goes on inside. It will be necessary to understand some of this 'wizardry' if you are to make the best choice when buying, and then to get the best out of it.

AMPLIFIERS WITH . . .

Amplifiers come in two basic breeds—either on their own, or built into something. If you buy a record player, for instance, which produces sounds all by itself when plugged into the mains and set in operation, then the amplifier is built in. The same goes for tape recorders and radios.

On the other hand, if your turntable will not produce the sound all by itself, then you need an amplifier.

AMPLIFIERS WITHOUT . . .

On their own, amplifiers tend to look uninteresting. They are merely boxes with knobs on. Certainly, they might be dressed up a bit to give an impression of elegance, but do not make the mistake of thinking that the most beautiful looking amplifier is the best. Be assured, it does not necessarily follow. Some of the most glamorous-looking models produce the most indifferent performances.

On the outside then an average amplifier has the following controls:

Volume	Filter	Radio
Bass	Balance	Tape Deck(s)
Treble	Turntable	On/Off

There may also be windows with needles and scales to show input levels, and neon lights which indicate when the amplifier is switched on.

CONTROLS IN PERSPECTIVE

By way of brief preliminary explanation, inasmuch as you will have a turntable, a tuner and a tape deck linked to the same amplifier, you will have to be able to control them. By switching to any one of the three or more (there may be provision for more than one tape deck), you can select which one will emit the sound signals intended to drive your speakers.

When you choose your amplifier, you should keep the future in mind. Alright, at the moment of purchase, all you have is a turntable. But already you are dreaming of having a tape deck and a tuner. So, think ahead and buy your amplifier with facilities for linking these in. Make sure it has the controls to handle all you want to add. If you make the mistake of not getting the right thing in the beginning, you are going to have to buy again when you are in a position to get new units.

FILTER CONTROL

The Volume, Bass and Treble controls are too familiar to warrant any explanation in depth, but the Filter and Balance may cause you trouble.

All modern amplifiers usually have some kind of filtering networks so that you can control bass and treble tones. As this type of circuitry is difficult to design for amplifiers, some knobs and switches on amplifiers in the cheaper price bracket are hardly worth having.

However, filtering does help. In simple terms, filtering is designed to reduce unwanted signals and to increase wanted ones.

Let us assume that you have turned up the volume of your amplifier to listen to a quiet passage of music. Because the record is a favourite of yours and it has spent a good deal of its life under the stylus, you are likely to pick up annoying crackles and pops (just like the breakfast cereal!) and other attendant noises. Naturally, your enjoyment of the music would be impaired.

In these circumstances, it would be nice to have a control that would subdue these extraneous noises. Surface noises such as these are usually in the high frequency range. Therefore, if something could be done to prevent such high frequencies from reaching the speakers, the resultant sounds would be of a very much better quality.

This is where the Filter control comes in. Its proper definition is a 'Low Pass Filter', and its function is to modify the way the amplifier deals with higher frequencies. There is also a filter which eliminates unwanted signals in the low frequency range. It is known as a 'High Pass Filter'.

The effect of these filters is usually quoted in the specifications as 'dB per octave'. This is a musical octave, and the specification shows how much attenuation (reduction) of the higher (or lower) frequencies takes place.

It is possible to get some filtering with the Treble control, but this is generally not as satisfactory as having a built-in filter control.

EQUAL SOUND

Finally, we come to Balance. When someone talks about Hi-Fi, he is generally speaking at least about stereo, if not in this day and age, about Quadraphonic sound. However, since Quad is so new and is still very much a status symbol, you will know right from the beginning if that someone is telling you about his all-round sound. The word 'quad' should be somewhere in the first sentence.

Stereo or Quad, the speakers have to be balanced so that each is working at the same strength. You can achieve this by rotating or sliding the Balance control.

TAPE MONITOR

When you are choosing your amplifier, you may or may not want the facility offered by some of being able to connect a tape recorder to a 'Tape Monitor' input—controlled by a Tape Monitor on/off switch on the front panel. When activated, this has the effect of bypassing some circuitry and feeding directly into the amplifier just before the output stages.

The advantage is that some noise and distortion will be eliminated which might be present on some of the less sophisticated amplifiers.

There is a disadvantage, however. There is no way of adding any correction through the Bass and Treble controls when the tape recorder is connected in this way.

THE HEART OF THE HEART

The average amplifier can be divided into three sections internally—the power supply, the pre-amp, and the power-amp. The power supply is what makes everything work inside the amplifier. In the pre-amp section are found all the inputs and controls. It accepts the signal and gets it ready for the power-amp, which decides the level of the ultimate sound signal, and the range (from very quiet to loud).

The only real way to assess the performance of the amplifier of your choice, or any other unit in your Hi-Fi set-up, other than listening to it, is to study its 'form'—that is, its specifications.

Now, hold on. Don't close the book. It is true that we said we would not foist any unnecessary technical jargon on you. Even though it looks as though it is about to happen, rest assured that you will only be bothered by that

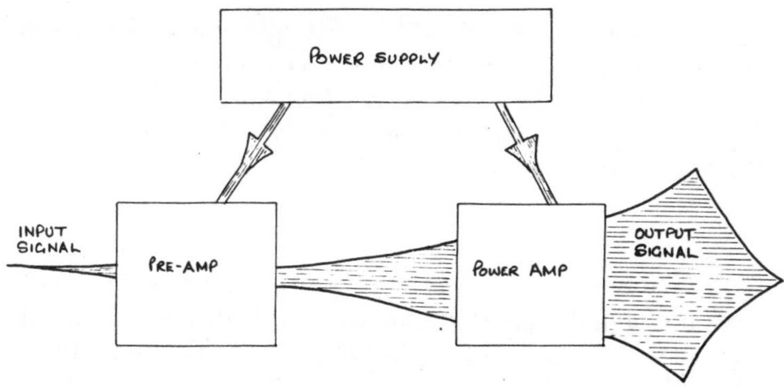

Fig. 9 Signal behaviour in an amplifier.

which is important in helping you to make your choice, and to aid you in your quest for optimum results.

However, all the way through this, retain one important thought in your head. Let your ear be your guide in the final analysis. Do not become a slave to electronic symbols and equations. Make them work for you to find out what you need, then let your ears be the final judge.

If the sounds which are boosted by the amplifier of your choice are pleasing to your ear, then you are on the right track. Remember that.

UNDERSTANDING THE SYMBOLS

O.K., let's begin. Take a deep breath and look at this little lot:

Power supply:	AC 210–250 V. 50–60 Hz
Power output:	20W rms per channel into 8 Ω load
Speaker matching:	6–16Ω
Distortion:	Total harmonic distortion, less than 0.1% at full output at 1 kHz
Frequency response:	20 Hz–20 kHz ± 1 dB
Signal to noise ratio:	Better than 60 dB below full output weighted.
Cross talk:	45 dB at 1 kHz

INPUT	*Sensitivity*	*Matching*	*Response*
Pick-up (magnetic):	4 mV	47 kΩ	RIAA compensated
Pick-up (ceramic):	40 mV	470 kΩ	Flat
Radio:	250 mV	470 kΩ	Flat
Tape:	250 mV	470 kΩ	Flat
Tape record outlet:	250 mV	470 kΩ	Flat
Overload capacity:	25 dB		

Now this is the kind of thing that you will read in a test report, or the instruction book of a typical amplifier. Daunting, isn't it?

Or is it? Look again, only this time, look closer. There are some recognisable symbols there. It does not matter at this point what they mean, just pick out those terms you recognise.

Seeing is believing. Let us see. The first thing you spot is 'AC'. Now that is easy. Alternating current—the electricity which comes from the mains as opposed to DC (direct current) which comes from batteries. Now you know that it is a mains only unit.

Then you see Hz. This is familiar too. You learned about Hz, or hertz, in the first chapter. It is the equivalent of cycles per second, which is the frequency of sound waves. Next is 'W'—watts, the symbol for the unit of power. But then you see 'rms'. That is a new one. Pass it for now because we shall explore it later.

Next is an old friend. Ohms (Ω) are something we learned about at school—and then we promptly forgot what the word meant when we left unless, of course, you entered a career in electronics or electrics. Yes, an old friend whose name does not escape us, but his precise nature does. Never mind, we will return to him later.

Note that after the second mention of ohms, we return to Hz and kHz, and then come face to face with dB (decibels, remember?). The next unusual symbol is mV. There are no prizes for guessing that it means 'millivolts', which are thousandths of volts. 'kΩ' (kilohms) are easy, but what is

'RIAA compensated'? or 'Flat'? These are new terms which need some explanation.

However, just stop now and tot up your score. On closer inspection you realised that you knew a great deal more than you had at first thought. You have become aware, in these few pages you have read so far, of the meanings of quite a few electronic symbols. But, it is not the meanings which are bothering you, it is the relationship they have with one another and into what context they are placed.

Let us see if this confusion can be cleared up.

SPECIFICATIONS STEP BY STEP

Power supply. Mains, as you have already worked out. But note that electric voltage can also be expressed in hertz, or cycles per second.

Power output. This, of course, is the ultimate power the Hi–Fi outfit in general, and in this case the amplifier in particular, generates after receiving an input of 250 volts.

You will note that it is expressed in watts rms into an ohms load. To begin to understand this, let us take a quick journey into basic electricity.

In a circuit you have a number of elements. To begin with, there is a power source which is usually either a battery, or the mains. The power, in terms of volts, is then activated by the operation of a switch. The result is called a current flow, and that is measured in amperes (amps).

Now, you may want to control the strength of the current, or produce heat, or even develop the voltage. To achieve this, you will need a resistor. Resistance is measured in our old friend, the 'ohm'.

Watts and rms. Ordinarily, to discover the power output in terms of watts, it is necessary to multiply the power (volts) by the current (amps)—provided the circuit only has a resistance. But, there is a third factor which must be taken into account.

Because an alternating current 'ebbs and flows', a power build-up occurs and the induced current begins to

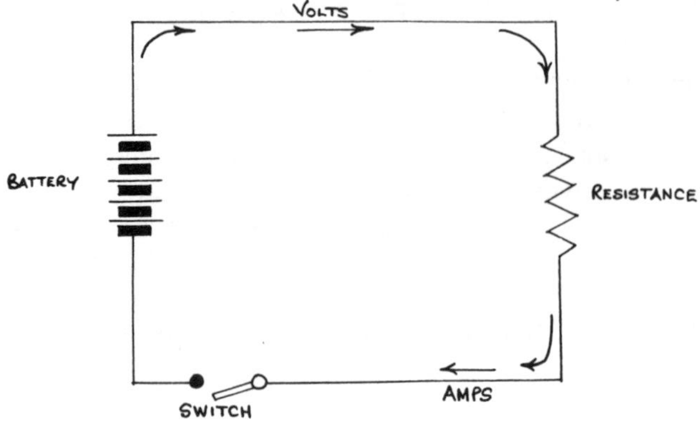

Fig. 10 Simple circuit.

oppose the applied current. The result? The amps get out of step with the volts.

This is called reactance, which acts as a form of resistance, and is measured in ohms too.

Now, the combined effect of resistance and reactance has another name—impedance. This is also measured in ohms, and must be taken into account when calculating power output in terms of 'watts'. This time, the formula is the voltage squared, divided by the impedance.

Watts rms can loosely be described as the method of finding the true average power output. The letters 'rms' mean 'root mean square'. To arrive at the rms, the experts take the average of the squares of all points between the peak and lowest point of a sound wave for one complete cycle, and then work out the square root of the final figure.

So, we have a power output from our amplifier of 20 watts rms from each stereo channel into a load (the element in a circuit to which energy is supplied) of 8 Ω—in other words, the impedance of the speakers.

The damping factor. There is another important aspect of speaker impedance which is directly related to amplifier impedance. The difference between these two is known as the damping factor. This determines how frequencies will respond. A low source impedance is desirable in order to

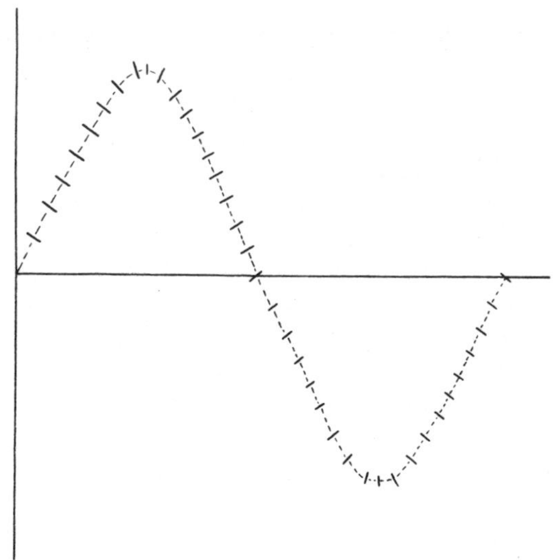

Fig. 11 Plotting the root mean square of a signal voltage.

produce a high damping factor. The higher it is, the better and more faithfully will the low frequencies reproduce.

However, the damping factor is becoming less and less of a problem. On today's amplifiers, modern circuit design is ensuring adequate damping factors, decreasing its importance as a specification for matching.

Speaker matching. This is an outline for a *stereo* amplifier, and it is designed to drive two speakers. In order to give a balanced sound, they need to be identical in their performance. Speakers are identified by ohms, and the figures tell us that we can plug speakers into this amplifier anywhere in the range suggested—as long as the speakers are identical. However, as the 'Power Output' specification indicates, the optimum speakers value to work with this amplifier is 8 Ω.

Distortion. Harmonic distortion is the form mentioned here. Every electronic unit, like musical instruments, has its own peculiar tones and overtones. Harmonic distortion occurs when undesirable overtones intrude. It is noticeable that the specifications of this model records a value of only 0.1% at full output at a frequency of 1 kHz. This will be different at other frequencies, but if you see a basic value of more than 2%, beware.

Frequency response. In simple terms, this specification indicates the range of frequencies to which the unit in question will respond. However, it is a little more complicated than that, because we also want to know how the unit responds to these frequencies. We can do that by putting frequencies together with dB to create a specific picture of the unit's potential performance in graph form.

The base, or horizontal line, is shown in frequencies, and the vertical line in decibels. The basis for the average response is 0 dB and is indicated by a line drawn parallel to the base about halfway down the graph. The resultant curve indicating how the unit reacts to different frequencies is then drawn. Whenever it drops below the 0 dB line, it is a minus figure, anytime it rises above, it is a plus figure.

Now we are well on our way to unravelling the term 'frequency response'. We know that our amplifier will handle frequencies from 20 Hz (20 cycles per second) to 20 kHz (20,000 cycles per second). This represents a reasonable range. For example, you will remember that the human voice, on average, has a range of 40 Hz to 10 kHz, and a symphony orchestra ranges from about 30 Hz to about 15 kHz.

If, however, we want to know how this frequency range of 20 Hz to 20 kHz reacts in our amplifier we must refer to our curve, which is known as the 'frequency response characteristic'. In the most ideal circumstances, it is expected that the main part of the 'curve' should be flat—that is, running parallel with the 0 dB reference line.

It is a fact though, that no Hi-Fi unit is perfect. Amplifiers are no exception. Even though they have a good frequency range, they still tend to fluctuate, especially at the extremes. This fluctuation can be seen on our amplifier as a figure of ± 1 dB. It suggests that over the frequency

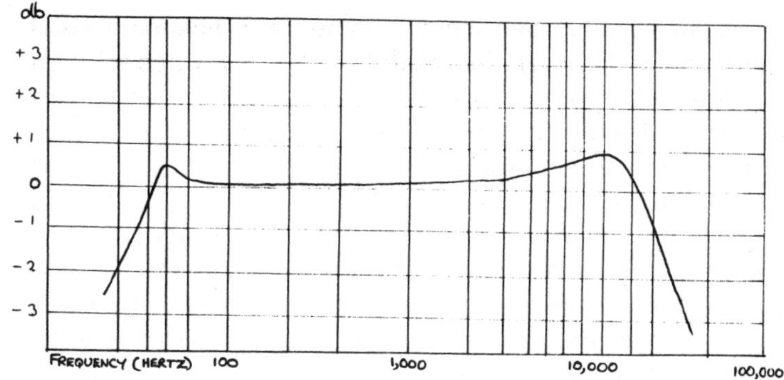

range the sound intensity will not fluctuate more than 1 dB either side of the curve line—viz. +1 dB to ÷ 1 dB, making a total of 2 dB in all. In real terms, you will not notice the difference.

Most modern amplifiers are made to respond to within ±1 dB. Some are slightly greater and can stand a certain amount of fluctuation without becoming noticeable. However, if the drop at the extremes is too great, some of the high and low frequencies will disappear, impairing the resultant sound.

Signal to noise ratio. This is the difference between the sound of the signal and the sound of the background noise caused by the random flow of electrons within the circuits. Note that, in this case, it is better than 60 dB below full output—in other words, it is virtually imperceptible.

Sometimes, you will encounter figures that have the word 'weighted' or 'unweighted' beside them. Weighted figures tend to be more helpful inasmuch as they are logarithmic measurements designed to put the unit specification on a par with the reaction of the human ear. Weighted figures will often appear to be lower and, therefore, better than unweighted ones.

Manufacturers will quote weighted figures, but will not mention the fact. This makes them appear to be better than the unweighted figure of another unit, but that may not be the case. The only safe way of comparing measurements is weighted figure with weighted figure, or unweighted with unweighted.

Cross talk. The transmission of recorded sounds from one stereo channel to another. Here it is recorded as a 45dB difference. However, since noise levels vary according to the frequency of that noise, 'cross talk' will rise and fall in volume too. Therefore, it is necessary to discover 'cross talk' levels at a variety of frequencies in order for it to be of any real value.

Pick-up/radio/tape. These three, *per se*, do not require too much explanation. We all know what they are. But, what this does indicate is, that inasmuch as they are shown in the specifications, the amplifier must have provision to link them up.

Nevertheless, the specifications themselves, and what they mean, do need to have some light thrown on them.

Sensitivity. This refers to the voltage input, and at the same time provides one factor in the information required to ensure that the unit you link up to the amplifier is electronically matched.

Notice how the input level is so tiny. It is measured in millivolts (thousandths of a volt). These tiny figures are common in electronics and demonstrate the incredible job that is done by the amplifier. Simply compare them with the output figures to see for yourself.

Please note that the symbol for 'milli' is a small 'm'. You will also encounter a capital 'M' often as you study Hi-Fi related electronics. This stands for 'mega', and has a meaning of one or more million.

Matching. Impedance is another word you will come across often too. You have already met it several times, and here it is again. The figures in this column are stated in kΩ (thousands of ohms). They are impedance (resistance and inductance combined) values which control the input voltage to the amplifier and are important factors in ensuring the units to which they refer are electronically compatible with the amplifier.

Power response. This relates to frequency response, which is, as you will remember, the relation of sound levels to frequencies, illustrated on a graph. The resultant curve on the graph demonstrates how the frequencies behave in intensity over the range the unit is capable of handling. The curve itself is known as the characteristic.

Looking down the column, we understand that 'flat' refers to the main line in the 'curve'. Ideally, the frequency characteristic line should run parallel with the base line to show that sound levels at all frequencies are compensated for and respond equally.

But, what of 'RIAA compensated'?

At one time, there was no standard restricting the percentage of frequencies used on records. For instance, the amount of room taken up by the bass frequencies sent many a Hi-Fi enthusiast running to his Hi-Fi outfit to make adjustments to compensate.

Then an organisation called the Recording Industry Association of America stepped in and established a standard to 'equalise' recording practices.

Having such a standard has allowed equipment manufacturers a framework to devise and build-in frequency compensation networks. Amplifiers are now able to exercise greater control over the sound signal coming from the record. It is achieved by amplifying some frequencies more than others to give a more even response.

Tape record outlet. When your tape deck is plugged into your Hi-Fi system, you will expect to be able to record direct through your amplifier from the tuner or turntable, without having to hold the microphone to the speaker. This means that you need a recording 'outlet'.

In modern amplifiers and the plugs which fit them, the input and outlet for the tape deck is channelled through the same socket and plug. Therefore, when recording, you merely play the turntable or tuner in the usual way, with the switch turned to the unit you are playing. Then, by depressing the record button, you will be able to transfer those sounds you can hear through the speaker, onto the tape.

Overload capacity. An amplifier must have a wide overload margin. The principle of 'overloading' can be explained by using the analogy of pouring water into a cup. If you pour it in at a steady rate, you can fill the cup to the brim without any problem. But, if it is poured in too suddenly and fiercely, it will simply splash back out of the cup and you will end up with more water on the table than in the cup.

Your amplifier is much like the cup. It must be so constructed to be able to accept a sudden high output from—say—a musical crescendo on a record, without sound 'spillage', or distortion.

In the days when amplifiers were made with valves, overloading factors were not such a problem. But transistorised amplifiers are a different kettle of fish. They are much more sensitive and have to have a wide overload margin built in.

This overload capacity must also be able to cope with signal intensity fluctuations from equipment changes. For instance, although most modern pick-up cartridges emit a similar output voltage, some emit more output voltage than others. Should you change your cartridge at some future date for a better one, it might produce more output. Obviously, you do not want to be forced into buying another amplifier so that it can cope with the higher voltage. A very important reason why a wide overload margin must be built into your amplifier.

Generating voltage. While most high quality cartridges have a registered voltage output of anything between 1 mV and 5 mV, the ultimate voltage generated can be considerably higher than this.

Voltage is generated as the stylus 'tracks' through the groove of the record. The voltage fluctuates in relation to the sound level on the record. This can be very much higher than the stated input level on the specifications. In fact, due to the high sound levels on records, the cartridge input of the amplifier should be able to handle at least 30 dB overload in relation to the input sensitivity.

Hot rodding. It is easy to fall into the trap of believing that because one amplifier has a better quoted overload margin than another, it is a better buy. This is not necessarily so.

Overload margins can only be defined clearly by looking at the overload point—measured in decibels—in relation to the input—measured in millivolts. This rule of comparing one with another applies to just about every other section in the specification list. For instance, an amplifier boasting a good signal to noise ratio will only tell you the full story if you know how much signal the amplifier has to cope with.

Without such comparative information, it is rather like a friend telling you that he has a car which reaches 100 mph. You would probably pass it off as nothing extraordinary. But, if he then said that his car was a 1958 Ford Popular, this added information would throw a different light on the first statement. You would then realise that he had a car with something special under the bonnet.

Similarly, Hi-Fi figures must show the whole picture to allow you to evaluate the potential.

INPUT/OVERLOAD RATIOS

Let us bring this back to amplifiers by looking at two different models and studying the overload margins. Number one amplifier is quoted as having an overload point of 28 dB. Amplifier number two is the model we have used as an example, and has an overload point of 25 dB.

You could be forgiven for assuming that amplifier number one is better than the second one, because, on the face of it, that is what the figures on their own say. But, when we study these overload points together with the input figures a different picture will emerge.

The first amplifier has a cartridge sensitivity of 2 mV, and number two, our amplifier, has an input sensitivity of 4 mV. Our calculation then becomes the input sensitivity multiplied by the logarithm of the overload point. As this is a voltage ratio, the decibels are considered to be 1/20th of a bel, not 1/10th.

Therefore, we have for amplifier number one the calculation $10^{1.4} \times 2$, which tells us that an input of 50 mV and over will cause overload. Our own amplifier's calculation tells us that an input of more than $10^{2.25} \times 4 = 36$ mV will overload.

So, now we see the truth. Number one amplifier, despite its apparently higher overload point, will go into overload much sooner than amplifier number two. With no apologies for repeating ourselves, the important lesson derived from this exercise is that you must compare one specification figure with another in order to paint the true picture implied by all those numbers.

KEEP CALM

Do not let the specifications of amplifiers frighten you. As you have seen, they can be 'licked', and all those figures and letters can be very useful in helping you to obtain an impression of the value of the unit in which you are interested. Those you see in Hi-Fi magazines may appear to be more complicated, but simply do what you did here—look down the list for 'friends'—and the picture will begin to take shape.

All technical editors do is test the same aspect of an amplifier (or other unit) several times, under varying circumstances, at different frequencies, and with different methods. Then they list all their findings. Just look one over and see. We have merely introduced you to a basic list by way of explanation and to give you a grounding in understanding amplifiers.

But, let us move on from this point. With the basic amplifier and its technicalities under our belt, so to speak, we will now have a look at other units, beginning with one which can be an integral part of an amplifier—the tuner.

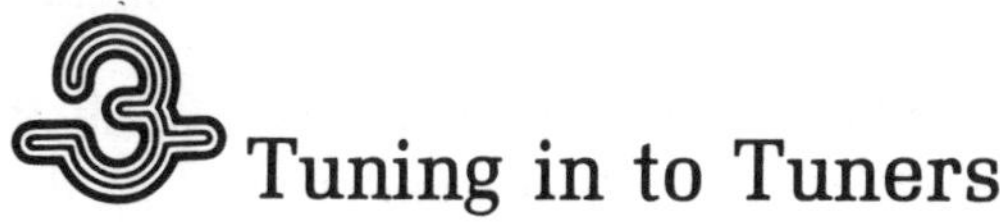

Tuning in to Tuners

The tuner will probably be way down your first Hi-Fi shopping list, but because they are often built together with amplifiers—known as tuner-amps—you may want to consider one as a possible 'first buy', rather than acquire them separately.

If this is the case, then you have a double job to do. You will have to consider the merits of the amplifier and the tuner section together to be sure of getting adequate quality and facilities from both.

PROS AND CONS

As far as the reproduction of high quality sound is concerned, combining the two makes no difference whatsoever. You can get just as good a sound out of a tuner-amp as you can from two separate units. The only real advantage to be gained from a tuner-amp is the space-saving facilities. One unit obviously takes up less space than two.

However, a combined unit does have a major disadvantage. The time will inevitably arrive when you will want to up-date your equipment. You can do it piecemeal—unit by unit—but a combined tuner-amp restricts you somewhat.

It can involve you in considerable, unnecessary expense. If you want a higher powered amplifier, for instance, you will have to trade in your integrated tuner as well.

With separates, your tuner, which may be perfectly adequate, can remain with you for years. On the other hand, your amplifier can be exchanged for the higher powered model of your dreams.

While no one actually considers future changes when at the first buying stage, it is worth bearing in mind at the point of making a decision about what to buy. Whatever your choice, we will take this opportunity to explore tuners in general whether an integral part of the amplifier, or not.

MODULATING SOUND WAVES

On looking at tuner merits there are certain factors which must be taken into account before you begin to delve into the electronic 'innards'. To begin with, you need to decide what programmes you want to receive. Secondly, you will have to discover what you can receive, and with what quality.

Let us begin with the first point. If the sum total of your receiving ambitions is three of the four BBC stations, and perhaps one or two of the local ones, then all you need is an FM (frequency modulated) tuner. If, however, you want to extend beyond that, then you ought to be looking at AM/FM tuners. 'AM' means 'amplitude modulated', incidently.

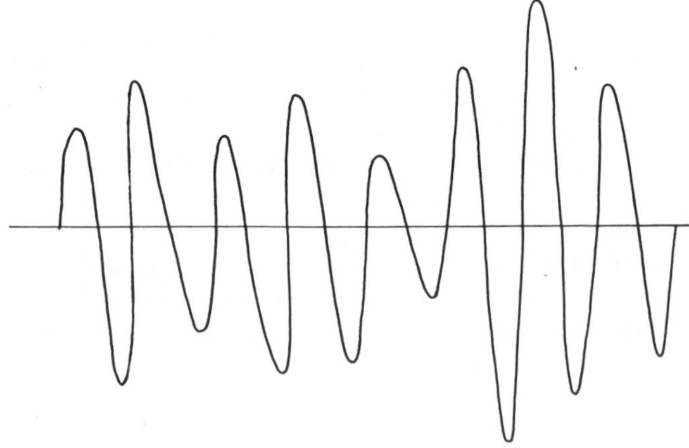

Fig. 13 Amplitude modulation.

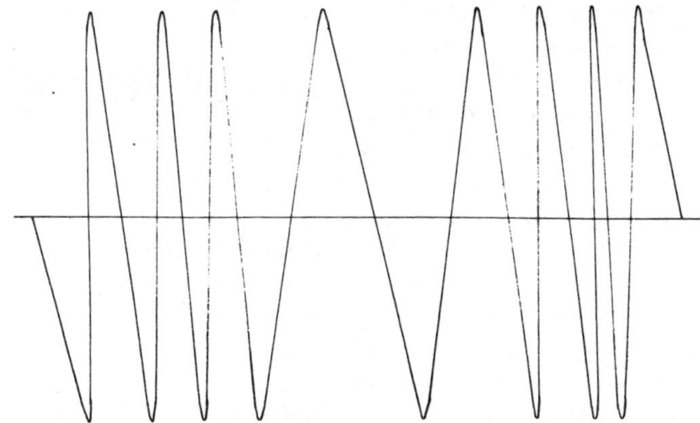

Fig. 14 Frequency modulation.

Let us get the technical terms out of the way. In a nutshell, radio waves cannot travel too far, so, they need a helping hand. That help comes from a higher frequency radio wave called a 'carrier'. The two are combined, and made to conform with each other, technically termed 'modulation'.

Now, there are two courses open to modulated waves. The first course is to maintain a constant radio wave frequency and alter the amplitude of the carrier to conform. The result of the two working together in this way is amplitude modulation (AM). Alternatively, when adding the radio and sound waves together, you can vary the radio frequency and maintain a constant amplitude. The result of this course is frequency modulation.

HOW RADIO WAVES TRAVEL

We will now step back a pace or two and have a look at how the basic radio wave is formed and how it behaves.

A programme is broadcast via a transmitter which is made up of banks of very large amplifiers connected to the transmission aerial. These contain enormous valves which dissipate high power at frequencies above the human hearing capabilities. These signals are transmitted as waves which travel at ground level. Not surprisingly, they are called 'ground waves'.

There are two types of sound waves transmitted in this manner—long waves and medium waves. Given an equal transmitting power, the long wave has the better effective distance.

Short waves. The third type of transmission—short waves—on the other hand, space-hop. They make use of the reflective properties of the ionosphere. The signals are directed at the ionosphere, reflected back to earth, re-reflected back to the ionosphere, to earth, and so on.

This makes short waves admirable for long distance transmissions. Extensive distances can be reached by this method, especially at night, when the ionosphere is at its strongest. That is why spies always transmit to their home base at night. But, of course, you know that. You see it on the movies all the time.

Interference. Night time transmissions of medium and long waves are not so good. You will have noticed that programmes are often subject to fading at night. This is due to the arrival at the radio set of a signal direct from the transmitter, followed microseconds later by the same signal reflected by the ionosphere. The delay of the second signal, because it had to travel some distance, makes the two arrivals slightly out of phase. The result? They tend to cancel each other out, creating a momentary loss of programme.

In addition to that, the medium waves are very crowded, and it is not at all uncommon to find two or more frequencies (stations) overlapping—giving you an unwanted two programmes for the price of one. This situation is amplified at night, again, thanks to the ionosphere, which enables these waves to travel even further than they were meant to.

Very high frequency. It is the lower wave lengths which operate using amplitude modulation—that is, the long, medium and short wave bands. The newer FM wavebands operate with very high frequencies (VHF), giving much clearer reception. Interference and background noises are virtually non-existent. However, because the ionosphere absorbs VHF waves, FM transmissions can only travel short distances (approximately 50 miles).

This means that it is necessary for you to be very close to a VHF transmitter in order to receive its signal. Fortunately, all the new stations have been built to transmit very high frequencies, so there are not many areas in the country which are now out of range.

Their short range does have one major advantage though, it lessens the chance of one VHF transmission interfering with another.

AM OR FM?

It would be superfluous to now go into detail about how you can only really get quality sound from an FM tuner. The fact that it is free from interference is a vital factor to take into account when making your choice. In addition, as stereo reception is virtually synonymous with modern Hi-Fi, the interference-free FM waves seem to score yet again.

But, do not discount AM. Yes, you will experience indifferent reception at times. No, the sound signal will not be as clean as those in the FM wave band—nor in stereo. But, AM will extend your programme choice.

Therefore, look at AM/FM tuners very carefully. The two together will give you everything you want—clarity and a stereo signal on the one hand, and a very much wider choice of programmes on the other.

But, wait! We have not finished with tuners yet. The radio waves are only a part of the story. We have yet to discover what makes one tuner better than the other, and how it works.

MORE SPECIFICATIONS

Let us follow the pattern established in the chapter on amplifiers and explore some specifications of a typical

AM/FM tuner. When you see them, they will look something like this:

Frequency range/coverage:	FM 88–108 MHz AM 525–1650 kHz
Antenna input impedance:	300Ω (balanced) 75Ω (unbalanced)
IHF usable sensitivity:	FM 1.8 microvolts (μV) AM 20 V mW; 40 V 1W
Output:	0.1 V into 10 kΩ
Harmonic distortion:	Mono 0.3% } 400 Hz 100% Stereo 0.5% }
Frequency response:	20 Hz–15 kHz ± 1 dB
Signal to noise ratio:	Better than 65 dB
Stereo separation:	Better than 35 dB
Capture ratio:	Less than 1.5 dB
IF rejection:	80 dB

Go through the same routine you followed with the amplifier specifications and see how much you recognise. Much more will be clear to you this time because of what we have already talked about. There are lots of familiar terms in this list. The meanings are more or less the same, and provided you have absorbed the information given so far, you will have little trouble.

SPECIFICATIONS ONE BY ONE

But, let us not dilly-dally. We must move on. As before, we will take each one of the terms in turn and explore their meanings.

Frequency range/coverage. This is an easy one. It refers to the frequencies of the radio waves received and handled by the tuner. It is the only guide to the stations you will be able to hear. Therefore, if you have particular programmes in mind which you would like to listen to, be sure you know the frequencies on which they are transmitted. Then, check the coverage offered by the tuner you have in mind.

In practice, you will find that all modern AM/FM tuners receive a very wide range of frequencies. Most BBC (national and local) and independent radio stations broadcast on FM, and the others are well and truly inside the AM wave bands.

Should you want short wave though, you will have to think again. You will probably have to consider a completely different set-up, divorced from your Hi-Fi system. This, then, really puts it outside the scope of this book.

Meanwhile, back at the VHF, Long and Medium wave bands, you can see that frequency specifications are in the old familiar 'hertz'. The FM wave band ranges from 88 to 108 MHz (1 megahertz = 1 million hertz), and the AM (Long and Medium) wave bands cover 525–1650 kHz (1 kilohertz = 1 thousand hertz). It is easy to see from these figures that FM is the very high frequency wave band.

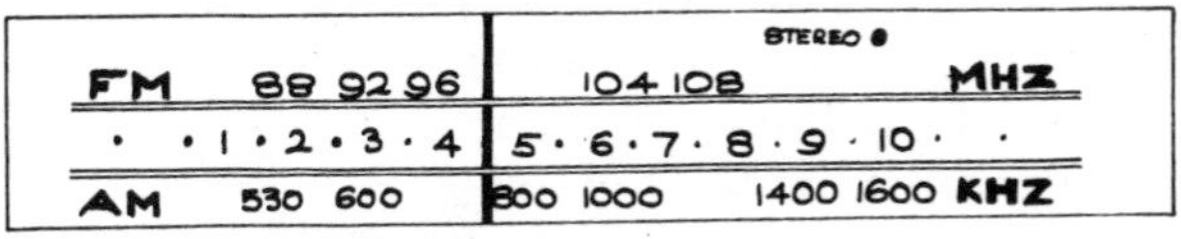

Fig. 15 Tuner dial.

Antenna input inpedance. The antenna is, of course, the aerial. All tuners need some sort of an aerial to improve reception, and consequently, performance.

If you have an AM/FM tuner, then you will need an aerial for each band. You will find sockets for both aerials in the back of the tuner. Fortunately, in most cases, you can get by with an indoor aerial, so the job is very much more simplified.

AM aerials. The requirements for the reception of AM frequencies are very basic. If you do not have one built into the set (unlikely, if it is a tuner), then you will only need a length of wire plugged into the appropriate socket in the tuner. With the aerial attached, you will notice a marked improvement in reception.

By all means, experiment with the length of the wire. There is an optimum length which will give optimum results from the AM band of your tuner. Do remember though, that the signal will not have the clarity of the FM signal.

FM aerials. FM tuners tend to require something special in aerials. Stereo reception is so critical. Even so, proper attention to aerials for FM reception is all too often neglected. In order to understand the implications, let us take another look at radio wave transmissions.

We learned earlier in this chapter that FM waves tend to be short distance travellers. That is because they are direct. They cannot rely on ionosphere reflections because they are absorbed. Therefore, if you are any distance away from the transmitter, your aerial must be positioned as high as possible.

The reason is quite simple. The earth is round and the transmitted signal is straight. The further away from the direct 'line of fire' you are situated, the higher your own receiving aerial must be.

For the best reception, you should fit an aerial constructed of aluminium which is conductive, light, and rust-resistant. Such aerials are not expensive. It is a modest outlay in comparison with the cost of a good tuner—which you have bought in order to be able to get good reception.

Aerials indoors. Fortunately, with the recent abundance of FM transmitting stations, you should be able to use an indoor aerial without any trouble. This is always provided that you live close enough to one, of course, or that you are not in an area of high interference. However, even with these advantages, you should make sure that you buy an aerial which is suitable for your area, and for stereo reception.

To give you an example, the so-called 'Half-Wave' aerial which looks very much like the type that was once used for 405 line TV reception is well suited for mono radio reception. But, it is not at all satisfactory for stereo.

Should you live in the shadow of a transmitter, then things will be different. It will receive in stereo. In normal circumstances though, FM stereo reception requires a lot more strength for the equivalent signal to noise ratio of mono. Therefore, an aerial of higher gain (signal ratio) is necessary if the programme is to be received in the quality it deserves.

Match-making again. In addition to all this, the aerial must match the input of the tuner properly for the optimum transfer of signal. Your dealer will know the area and will be able to advise you about which would be best to use. If you fit one which is not sensitive enough, you will receive a poor signal, and the quality of reception will be of a very low standard.

On the other hand, an aerial too sensitive for your tuner and your area, could overload the input with too much signal.

It is possible that you will be situated in an area where a perfectly adequate signal will be received by a length of untwisted flex. This is attached to the FM aerial terminal in the tuner and runs for about 2½ feet. It is then separated into two 'strands', spread out horizontally, and is either fixed to the wall, or even laid out on the floor.

Ohms and aerials. Our example in the illustrated specifications shows that the input matching is measured in ohms. As you may have guessed, it refers to impedance. With this tuner, you would need a 300 Ω aerial, unless you have a 'ground' antenna. In which case, you could use an aerial of 75 Ω impedance.

IHF usable sensitivity. This determines the sensitivity to weak signals. It is also an indication of the amount of noise distortion which stays on the signal after demodulation in the tuner. Therefore, it shows first the relative freedom of the tuner to internal noise when there are gaps in the modulation and, second, the relative freedom from distortion when modulation is at a maximum.

Sensitivity measurements. There has been, in the past, confusion over the measurements of sensitivity. This has arisen because a variety of methods have been employed to arrive at the answers.

Some manufacturers took measurements of the voltage needed across the aerial input terminals for a certain signal to noise ratio; others measured these voltages in open circuit conditions; still others measured them in closed circuit conditions. The result of all this was completely different values.

This was further made nonsense by the fact that these figures differed yet again between stereo and mono.

A standardised system. Fortunately, this has now become standardised. Rather like the RIAA in record equalisation, the American Institute of High Fidelity has devised a standard by which measurements are made.

The initials IHF indicate that the measurements have been made to a common standard and two tuners of different manufacture can be compared. If you do encounter specifications where IHF does not appear, it merely means that a different system could have been used. The danger is that the figures could look as though they are better, but they need not be. So, watch out for that.

Output. This is the voltage of the final signal after it has been converted by the tuner, and is then ready to be boosted yet again to an audible level by the amplifier.

Inasmuch as the two must work together, the tuner output must match the amplifier input level. It is not critical. For example, we have an output of 0.1 V, or 100 mV. To match, it should 'look into' an amplifier with a sensitivity of 100 mV and an impedance of 10 kΩ or above. Generally, there will be no matching problems if the units are of the same manufacture. But, as you are choosing separates to tailor-make the best quality sounds for you, it is highly possible that the tuner and the amplifier will come from different stables.

If that is the case, then the tuner may require some adjustments to make it match. Some have an output level control to make things easy. Those which do not should be altered by an engineer. Do not try to do it yourself.

Harmonic distortion. As you have read before, this is a definition of quality of sound, with reference to the percentage of distortion which occurs as a result of the intrusion of inherent overtones (harmonics).

You will note that the percentage figures (for mono and stereo) are shown in relation to a frequency (400 Hz) at 100% modulation. The harmonic distortion figure is determined to be a percentage of the 100% modulation signal at 400 Hz. The figure could, in fact, alter at another frequency.

It is significant that the harmonic distortion figure is a higher percentage for stereo than for mono. This is not to say that the mono signal is any better to listen to. In fact, at the levels in the specifications, you would not even notice the difference.

However, it is fair to say that the stereo signal is pushed through a lot of additional processes in order to reach your speakers in stereo form. It is natural that because of these, distortion will be marginally increased.

Frequency response. The type of information given here is identical to that offered in the amplifier specifications. The figures differ only slightly. Compare them and see.

Signal to noise ratio. Stay with Chapter Two if you do not remember the exact meaning. To refresh your memory, it is the difference between the sound signal and the background electronic noise.

Stereo separation. This has similarities to the signal to noise ratio in that it is the standard measurement of the ratios of two different signals. In this case, it refers to the level of the unwanted signal 'leaking' into the other channel. The figure quoted, in dB, usually refers to the frequency of 1 kilohertz. The separation will alter at higher and lower frequencies.

Capture ratio. For FM reception only, it applies to the phenomenon that signals on any given frequency tend to suppress (or capture) weaker signals on the same frequency. They are not eliminated, merely reduced in volume. In this case, the reduction is less than 1.5 dB—in real hearing terms, virtually non-existent.

IF rejection. In the sections on 'Input' and 'Output' you learned about the function of a tuner to take a very tiny signal and amplify it considerably. The basic problem with achieving this feat is to maintain an output with an acceptable signal to noise ratio.

The problem can be overcome if the amplification is done at one frequency. Other, unwanted frequencies can then be reduced. This single frequency is known as the intermediate frequency (IF).

The receiver which uses this system is known as 'superheterodyne'.

Tune in to superhet. In very general terms, the system works like this. When you tune into a specific frequency to find the programme you want, the frequency received is mixed with another produced inside the set by an oscillator—a device for producing or converting power.

The frequency oscillated is equal to that already being received, plus 10.7 MHz. The result of this mixture is a final frequency of 10.7 MHz which incorporates all the 'sound' signals which were in the original.

Boosting the signal. This new frequency signal is now fed into another stage called the IF Strip and Detector. Here it is amplified, and any signals not based on the created 10.7 MHz are rejected.

Sounds familiar, doesn't it? Yes, indeed. We are really looking at IF rejection at this point. This is the volume of the ultimate sound difference between the frequency to which you have been tuned, and others which are rejected at this stage. In this case, the unwanted signals are 80 db below the desired frequency.

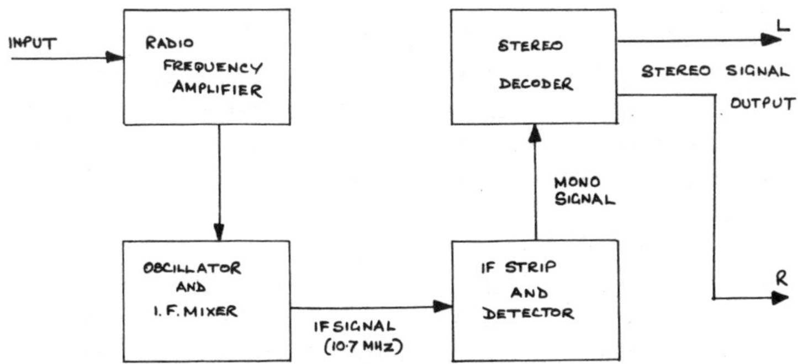

Fig. 16 Chart of radio signal progress in a superheterodyne receiver.

But, to continue following the course of the signal to the output stage. In the IF strip and detector, the signal is amplified to large enough proportions for the original sound signals to be extracted and transmitted as a monaural signal to the stereo decoder. The signal is then converted to stereo and continues to the output points.

From here, of course, it is channelled to the amplifier, boosted still more, and transmitted to the speakers.

FIND YOUR TUNER

Now, it is up to you. Read the specifications of any tuner or tuner-amp combined in order to whittle down your choice to manageable proportions. Hi-Fi magazines will give you plenty of leads. Study their recommendations and then go to your dealer to look the models over at first hand.

But, remember. Listen to them. Do not take anybody's word for it entirely. When you have pared your choice of tuners down to a few, let your ear help you find the one which you will enjoy.

4 Gramophones—Up to Date

The odds are that the turntable is the hardest working part of any Hi-Fi system. This means that besides being a precision piece of engineering it also has to be hard-wearing.

One of the most important criteria is that it runs almost constantly, for years and years, at pre-selected speeds—33⅓rpm, 45 rpm and 78 rpm (rpm = revolutions per minute). There can be no deviation. If the turntable rotates at 33½ rpm, or 46 rpm, or 77 rpm, the sound from the record will not be reproduced correctly—it will be off-key.

It is surprising how many Hi-Fi tyros consider turntable selection to be the least of their problems. It appears to be such a simple piece of equipment, which, as long as it works, does not need too much buying attention.

But that is where they are wrong.

SOUND SOURCE

The turntable is a vital link in your Hi-Fi chain and has to be good. You might describe it as a sound source. While your amplifier and speakers might be the finest pieces of equipment in the world, if your turntable does not turn out good original sounds, the faults will simply be amplified through the system.

You will hear them alright. Then, as time goes on they will become more and more acute to your ears. Finally, you will surrender to the inevitable and rush out to buy a better turntable.

To demonstrate the importance of the turntable, we will deal with the subject in this chapter in three parts:

Part 1. Creating a record
Part 2. The Turntable
Part 3. The Pick-up

You may think that Parts 2 and 3 belong to this chapter, but that Part 1 will not help you to better your ultimate choice of turntable. Perhaps not. But you will understand something of what the turntable has to cope with. Records are, in reality, inseparable from the turntable. They create the sounds that you want to hear. Therefore, it is worth taking a brief tour of the recording studios and a factory to see how they are made.

FINDING 'REAL' SOUND

The proper yardstick to use when buying a turntable is to choose one which does justice to the records. A great deal of work has been put into producing beautiful sounds which are precision-pressed into that piece of vinyl, and the object is to reproduce those sounds as faithfully as possible.

Do you remember the definition of Hi-Fi mentioned in Chapter One? It is the attempt to create the illusion that you are hearing the real thing.

That is how good your turntable must be.

At the very least, if you become aware of the work that is put into producing a record to make it sound 'real', then you will treat your turntable buying with utmost respect.

PART 1. CREATING A RECORD

Many years ago, recordings were made directly onto shellac records on a machine which looked like a more sophisticated dog-listening horn. The horn took the place of amplifiers and boosted the sound both ways—into the record, and out of it.

When recording, it picked up the sound coming from the source and funnelled it down to the cutting stylus. When being played back, the sound travelled the reverse route.

Then, after World War II, the tape recorder made its debut in the studios as the medium for recording sounds prior to them being transferred to disc. That same system is being used today. Of course, the equipment is much more sophisticated and the quality of sound reproduction is much improved. But the system of recording has not altered.

The quest for perfect sound goes on. No longer do groups of singers stand around a single microphone suspended from the ceiling. Now, each has his own microphone into which he sings or plays his heart out. It is not unusual to see anything from 20 to 24 mikes being used to record an orchestra, or a rock drummer surrounded by 10, or even 12 mikes.

The important thing in this quest for perfect sound is that each microphone must be so designed and so positioned that it only picks up sounds coming from the instrument or voice using it.

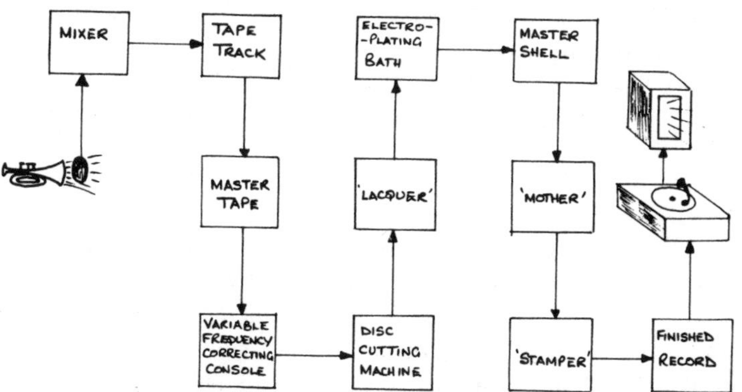

Fig. 17 The record-making process.

ISOLATING SOUND

In the case of a single vocalist, he or she occupies a small sound-proof room built within the studio. It is so designed that the orchestra can be seen, but not heard. It is acoustically isolated from the singer's booth. With this method of sound isolation, the sound engineer has more control over the respective sound levels of singer and orchestra or backing group.

For instance, if the singer and the orchestra were recorded together in the same sound area and he or she were situated in close proximity to—say—a loud section such as the brass, the weaker, human voice would be drowned out by the raucous brass sounds spilling into the voice mike. This makes good sound balancing an impossible task.

Move the microphone further away from the orchestra and another problem arises. The mike would still be able to pick up the other loud instruments in the orchestra. This background noise will then 'hover' around the voice and will sound distant, with a hollow quality instead of a warm presence and closeness which is needed for a rich orchestral sound.

Moreover, this sound spillage problem exists between different sections of the orchestra. There needs to be some acoustic isolation here too. Strings need to be separated from the brass, woodwind from percussion, and so on. It is, of course, not possible to put each instrument in a sound booth, but what can be done is to isolate sections by screens. Then each microphone will only pick up the sounds of each section.

MAKING A MIX

The signals received by all these different microphones are fed to a mixing console. This is a device which enables the signals to be treated in a manner designed to enhance that particular group of instruments or voices. Frequency corrections can be made to such an extent that the ultimate recorded sound can be better than the original.

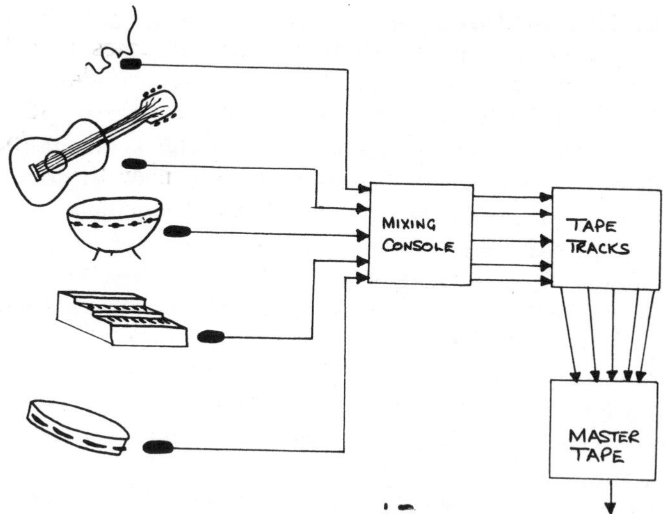

Fig. 18 Mixing musical sounds.

In some respects, this capacity of the mixing console is similar to, but more complex than, the bass and treble controls on a modern Hi-Fi outfit. However, the mixing console has an additional and more important function, to control and manipulate the individual sound levels on each microphone in order to maintain the correct balance between voices and instruments.

THE PLACE OF THE TAPE RECORDER

The sound from the orchestra and singer then passes through the microphone and mixer to the tape recorder. But this is no ordinary recorder. While it may perform the same function, it uses tapes some two inches wide.

These are multi-track machines 'with a vengeance' because they can record up to twenty-four tracks simultaneously, or one by one. Sounds collected by one microphone can be recorded on one track on its own and can be balanced against the string section on another track, and a singing voice on another, and so on.

What is more, the backing vocal group can be added to the lead singer's voice at a later date, or indeed, removed completely if so desired, and replaced with a different arrangement.

MULTI-TRACKING

A typical pop group of today might consist of about five musicians, all of whom are talented enough to be able to play several instruments. The multi-track machine allows different instruments to be recorded at different times, thus building up the effect of a full pop orchestral sound from just a few people—or in some cases, only a single musician.

This search for something different in recorded sound is undertaken by many modern musicians, making records an experience which cannot be repeated in the same way in real life.

Even so-called straight-forward recordings are built up piece by piece in the search for a perfection which cannot be repeated on stage. Sometimes entire scores are 'constructed' on record by playing musical phrases over and over again until the musician is satisfied. The best of these is then selected and added to the next phrase to make up the complete record.

MAKING THE MASTER TAPE

When all the tracks are satisfactorily recorded, they must eventually be transferred as one complete, mixed sound to a standard ¼-inch tape. All the tracks must be carefully balanced for the correct sound and these re-recording sessions sometimes demand the combined experience and expertise of the engineer, producer and artistes.

On their way to the ¼-inch tape, the 24 tracks are routed through the mixer again. Here they receive further electronic treatment if necessary, such as frequency correction, echo and reverberation, to improve various individual sounds.

The finished ¼-inch product is called the Master Tape.

CUTTING THE 'LACQUER'

This master tape is played back on a high quality tape

machine and passed through another variable frequency correcting console (just like a small microphone mixing console) to the Master Disc Cutting Machine.

This is really like a sophisticated record player, only the sound signals go the other way. Instead of being picked up by the stylus and 'channelled' through to the speakers, the sound signals are sent into a special stylus known as a cutting head. This diamond stylus 'cuts' the pattern of the sound signals into the groove of the blank record.

Now, this is no ordinary record. It is a thin disc of aluminium, about twelve inches in diameter and coated on both sides with soft vinyl. The official name given to the coating is the 'lacquer'.

During the 'cutting' process, to assist the head to make its mark, a controlled amount of heat is applied to the disc to soften the vinyl.

Only one side of the master record is cut. Side two, or the 'B' side, is put onto a separate disc. So, when finished, this aluminium disc is, in effect, a one-sided record—and, it can be played.

However, it is normally played only once on very high quality checking equipment to ensure the 'cut' is all that it should be. Repeated playings of this soft material would cause the quality of the sound 'cut' to deteriorate very quickly.

BALANCING THE SOUND QUALITY

We all like the resonance of deep bass notes emanating from our Hi-Fi outfits. Well, it might be fun for us, but for the record manufacturer, bass creates problems. Violent oscillations from bass frequencies cause groove distortions.

To prevent this happening, the cutters' amplifiers automatically attenuate an internationally agreed amount of bass frequency.

At the other extreme, to enhance the quality of the sounds emanating from the record when played, disc noise is reduced by purposely applying a pre-determined amount of treble boost.

The resultant problem is, the record has a lack of bass and too much treble. As the record at this point is the finished product, there is not too much that can be done here. The only place these compensations could be made to introduce bass and cut back treble is in the amplifier.

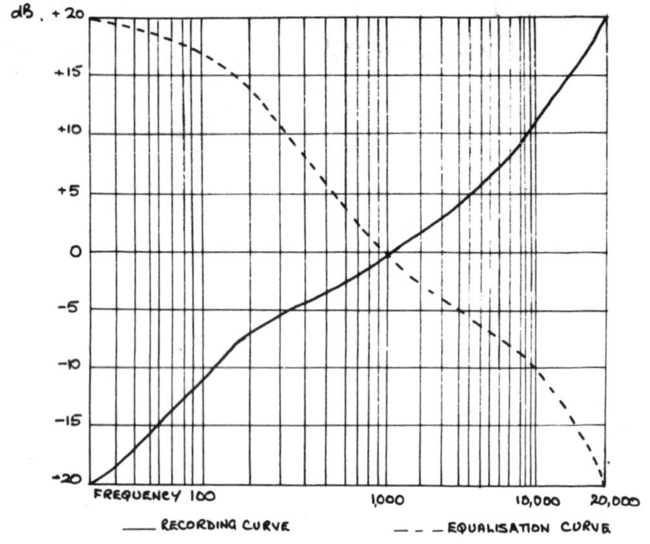

Fig. 19 Recording and equalisation curves.

EQUALISATION STANDARDS

In the early days of the gramophone record, it was necessary to incorporate into the reproduction amplifier, a selection of different equalisation characteristics. The amplifier had to handle a variety of records which were manufactured by different companies with equalisations which were all their own.

Things are different now, as you have already learned. A standard has been fixed by the Recording Industry Association of America (RIAA), and thanks to this, manufacturers of amplifiers have only one equalisation characteristic to deal with. The initials RIAA have now been adopted internationally to signify compliance with the standards laid down by this group.

OF 'NEGATIVES' AND 'MOTHERS'

When the master disc is complete it is ready for special treatment in the processing department. First, the 'lacquer' has to be washed and then immersed in a bath of chemicals until a silver, mirror surface is produced. This step makes the lacquer electro-conductive.

It is then transferred to an electroplating bath where nickel is coated onto the silvered surface. It soaks in this bath for about 3 hours whilst a current of about 120 amperes is applied.

As a result of all these baths, a solid coat, or what is known as a 'former', has been added to the lacquer. It is about 1/23rd of an inch thick and is called the Master Shell or Negative.

The master shell is lifted from the lacquer and, *voilá*, we have an inside out record—that is, a negative disc. It has ridges instead of grooves.

Then, this series of processes is repeated all over again, this time on the master shell. The impression we get from this is called the 'Mother', which to all intents and purposes is a one-sided record having a groove just like the discs you have in your collection.

PUTTING TWO SIDES TOGETHER

But wait—we have not finished yet. Still another plating process takes place, this time to create the 'Stamper' from which the finished records—the type you buy in the shops—are made. When that is done, it has to be 'formed' (trimmed) to fit the press. The reverse side of the stamper has to be smoothed off to prevent any blemishes being transferred through to the groove side. Finally, the surface must be cleaned and polished.

Sides 'A' and 'B' stampers are then fixed to the record press along with the pre-printed labels. Malleable vinyl is positioned in between the stampers and the press turns the heated material into the finished record.

PART 2. THE TURNTABLE

The next stage in the search for perfect sound is up to you and your choice of equipment. We will assume that you have selected a beautiful amplifier and now have to choose your turntable.

You will have noted the opening remarks and you no doubt have the desire to do justice to the records you play. So, you will not be cutting corners.

FIRST THINGS FIRST

However, even with this priority, the first criterion will be money. Let us face it, galloping inflation creates special kinds of problems for Hi-Fi enthusiasts. All you have to spend is change from the food bill! So, when you have counted it, the only thing you can do is look around for the turntable with a price that fits your pocket.

The trick, as always, is not to get the most expensive, the most beautiful machine you can afford, but to get value for the money you have available. You do not have to jump off the local church tower just because you cannot afford the turntable of your dreams. You will simply have to do the best you can. Despite what has been said, this is one unit on which you can economise to a certain extent without creating extreme sound problems. However (and we make no apologies for repeating it), do your utmost to get the best value you can for your hard-earned cash.

LISTEN TO THE DIFFERENCE

The second criterion is how the turntable sounds when it is in operation. By the time you have finished this chapter, you will know what to listen for besides the music. But, in the meantime, you might consider how it sounds to you.

As a newcomer in this field, you may find it difficult to detect any appreciable difference between the sounds emitted by the very expensive turntable and those by the less expensive machine. This is the time to remember that you are buying a Hi-Fi outfit to please you, not to impress your neighbour. So, if it sounds alright to you, that is a good first standard to go by.

In any event, there is always 'fail-safe'. When you are not so pressed for funds, and when you begin to recognise better quality sound, if your record deck does not come up to standard, you can always trade it in for something better.

One important point which you should take into account when listening is that the sound quality you hear in the shop is often very much better than at home. To reiterate something which has been said before, there are two reasons for this. First, the acoustics in the shop are generally very much better than in the average home. Normally, you do not set up and furnish your lounge with acoustics in mind. In most cases, it is the television which decides how the furniture should be positioned.

Secondly, you must consider the remainder of the equipment to which it is linked. Unless you specify it, and then only if he has it, your dealer is unlikely to plug the turntable to which you are listening into the same amplifier and speakers as those which you have at home. This can make a considerable difference, especially as he will deliberately choose those which function the best. Tip the odds in your favour by insisting that the remainder of the equipment used in the listening test is the same as that with which it will be used for the rest of posterity.

Remember to take your own record along too. It will provide a standard factor in the tests for sound quality. You will know that the same sound potential exists with each test, so it is just a matter of what the turntable does with it.

TURNTABLE SPECIFICATIONS

When shortlisting the turntables of your choice, the job is made very much easier by the fact that there are few specifications to wade through. When you think about it, there is little that can be said about the turntable itself—not counting, right now, the pick-up arm and the cartridge. They will be dealt with later.

All we really need to know is whether the platter turns at the desired speed all the time and whether the motor is hard-wearing and silent. Beyond this, there is very little to look for.

So, with that in mind, let us have a look at a typical turntable specification line-up.

Speeds: 33⅓, 45, 78 rpm.
Wow and flutter: 0.05% rms weighted.
Rumble: Better than −60 dB.

There now. You can cope with these can't you? Most of it is familiar to you anyway, but, we will dissect it to illustrate the relationship of one to the other.

Speeds. There is nothing irregular about this group of figures. The three standard rpm speeds—33⅓ (for long playing albums), 45 (for modern singles), and 78 (for singles of yesteryear)—have been around for a long time.

Things are beginning to change nowadays though. New turntables are being produced without the 78 rpm speed. This setting is being used less and less, especially as many recordings which have hitherto only been available on old 78s are now being released together with others on special albums. There are still enough collectors around however, to warrant including the setting on some machines.

The second major modification applies to those turntables which are fitted with automatic pick-up arms. You may find no less than three 33⅓rpm settings on the speed changing dial or lever. They are there for different sized records—the standard 12-inch album size, the smaller 10-inch LP, and the diminutive 7-inch record.

Getting into gear. It is interesting to see what goes on inside the turntable when you alter the speed to suit the record you want to play. There are several different methods used to achieve this change and as you read the details about the turntable of your choice, it is worthwhile noting which system it uses.

First, the most widely applied, and certainly the oldest method of making the turntable rotate, is the mains operated motor and rubber drive-wheel system. As you select the speed, a stepped capstan is brought into contact with the rubber drive wheel. This, in turn, is pressed against the inner rim of the turntable platter.

When the motor is switched on, the drive wheel begins to rotate, and as a result, causes the stepped capstan and the platter to turn. The speed at which the platter turns is then dependent on which 'step' the drive wheel is in contact with.

It is the rubber drive wheel which has to maintain the constancy of the speed, and it is the capstan which determines the correctness of the speed—i.e. the revolutions per minute. The capstan is adjusted by a system of interacting levers and arms and, each time the speed is altered, a different step is brought into contact with the spinning drive wheel. The smaller the step, the slower the platter speed.

This means that the central point of the platter has to ride evenly and without friction. You are already aware that this unit of your set-up has to undergo considerable use. Under these circumstances ordinary wear and tear from almost constant use has to be counteracted. As in many other instruments with moving parts, the ubiquitous ball-bearing helps to solve the problem.

Belt drive. A very popular type of turntable drive system uses a belt. As with the rubber drive wheel, described above, this system also makes use of a stepped capstan. But, this capstan has grooves channelled into it and is connected directly to the platter by a rubber band, or belt, seated in the grooves. In this case it is the capstan which turns directly from the motor and by means of the belt, drives the platter.

Infinitely variable speeds. Another method makes use of an elongated conical shaft instead of a stepped capstan. The motor is transverse, but it drives a rubber wheel which is in contact both with the cone and the underside of the platter. As the drive wheel rotates, it turns the platter at a speed commensurate with its position on the cone.

There is one distinct advantage with this system. The speed control is mechanically coupled to the drive wheel and thanks to the cone-shaped capstan, is infinitely variable. Nevertheless, the manufacturers of this type of turntable do provide click stop speed settings for normal record speeds.

Direct drive. This is probably the most advanced turntable drive system. The design is interesting. The centre spindle and the outer motor casing rotate together, therefore, the platter is driven directly by the motor. Drive wheels, belts and capstans are eliminated, reducing the number of moving parts.

Given a good, robust motor, made within precision tolerances, this type has distinct advantages. As already stated, there are less mechanical parts than in other types and, as a result, wear and tear is reduced. With friction so minimised, noise transmitted to the pick-up cartridge via the turntable is very slight.

Electronics and turntables. Speed stability in turntables has been the subject of much research. A number of different methods have been devised. One of the first was a motor which relied on the mains frequency (50 Hz) to drive its motor rather than the voltage. Known as the synchronous motor, it is not affected by voltage fluctuations and results in greater stability in turntable speed.

In another popular system, the rotation of the turntable platter is designed to supply voltage to a regulating circuit. In yet another, the voltage is sent to the motor from a regulated supply. Both types are able to iron out speed fluctuations in a split second.

The 'servo' system sees the introduction of electronics into turntable design. Here, the speed is monitored by a lamp and photo-electric cell arrangement.

Engraved into the outer casing of the motor are stroboscopic marks, similar to those on the rims of some turntable platters. A light is directed at the gradations and reflected into the electric eye. This, in turn, uses the rate of interruptions of the light reflections to 'drive' a circuit. Any speed deviation is then counteracted instantaneously by voltage correction.

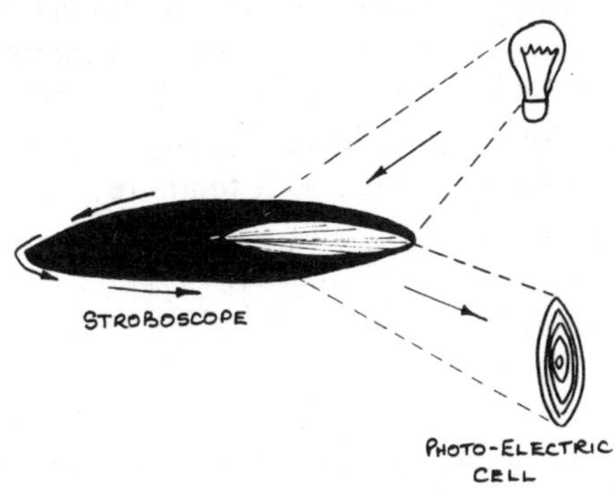

Fig. 20 Electronic servo speed control.

Now then, if fluctuations of power feed occur, either higher or lower than normal, the basic speed at which the motor runs will adjust automatically the moment the sensing device detects a change. However slight the deviation, the information is passed to the circuitry and will subtract or add the amount of voltage necessary to maintain a constant speed.

It all happens in microseconds and is completely undetectable by the ear. As all direct drive turntable speeds are electronically monitored, these units are much sought after.

Platter matters. Platters come in all shapes and sizes. The most common are circular discs upon which you position your record. But, modern technology is turning up some very odd shapes indeed. However, at the time of writing, the odd shapes are still very much in a minority.

Besides providing a smooth surface upon which to put records, platters can help maintain steady speeds. Minor fluctuations are ironed out by the weight of the platter, which can weigh as much as 7 lbs. These heavier platters are easily spotted because they tend to be very thick and have been given the name of 'Transcription Decks'.

While the heavy platter can be used with belt-driven motors with effect, it does create problems when put together with the electronically controlled motor. A turntable of this type has to be able to react instantaneously to voltage fluctuations, as we have already said, but a heavy platter causes inertia problems. It tends to carry on at the speed its weight is taking it and, as a result, takes time to adjust.

Therefore, the lighter platter is better to use in conjunction with your electronic motor.

Wow and flutter. These are terms which describe fluctuations in the speed the turntable platter is rotating at. You can probably imagine the effects, even if you have never heard them. The music being played begins to slow down, dropping in pitch. Then the pitch begins to rise, and rise, and rise, until it is too high with the music being played too fast. Then, it drops and starts all over again.

If it occurs relatively slowly, it is known as 'Wow' (it actually sounds like wow). If it happens fast, in a series of short fluttery fluctuations, it is termed as 'Flutter'.

Now, it is vital for you to check whether any audible wow and flutter is present before you buy your turntable. It might seem like a quaint little fault, but it is one you can certainly do without.

The only way to check of course, is to listen. But, listen carefully and closely. If the slightest degree of either of these two characteristics is present, move on friend, move on.

The figure in the specifications gives you some indication of the turntable's predilection for wow and flutter. It is expressed as a percentage of the average speed which is arrived at by calculating the root mean square. Our specimen turntable has a 0.05% rms wow and flutter factor. Any which have over 0.1% rms should be treated with utmost suspicion.

Rumble. What is it? A synonym for an earthquake perhaps? *West Side Story* fans are confident that it is a gang fight. But, no, it is neither of these things. It is merely audible turntable motor noise.

You see, as we have advocated all along, you need a good, relatively powerful amplifier. However, the problem is, a higher power amplifier—one of say, 20 to 30 watts rms output—will not only boost the sounds coming from the record, it will also amplify other noises in the Hi-Fi system.

Turntables are especially prone because of the motor, other moving parts (friction noises), and so on, especially in the cheaper models. This is made even more noticeable by the use of big speakers with an extended bass response.

The mechanical noise transmitted from the motor or drive assembly is picked up by the cartridge and passed on to the amplifier with the programme content. It can be heard then, most clearly during the quieter passages of music and comes over as a very low, continuous rumbling noise. Once you hear it, you will recognise it. Rumble is most distinctive.

RECORD CHANGING SYSTEMS

Before leaving Part 2 of this chapter, let us briefly survey the various methods of record changing available at the moment. Many Hi-Fi purists will not even consider record changing systems where automation is an integral part. They even go so far as to say that such systems are not really Hi-Fi.

Well, that may or may not be the case. The important thing is, you will be faced with making a choice from the systems available when you buy your turntable. Naturally, you need to be armed with the pros and cons.

What do you need? To begin with, your choice will have to be based upon what you want to listen to. If you are a singles fan, then the automatic changer is virtually a must. The only alternative is to pay continuous attention to your turntable while you are playing records. This will probably be far short of convenient most of the time.

If, on the other hand, you intend predominantly to play albums, then we would strongly advise against automatic changers of any decription.

Records stacked one on top of the other, while highly desirable for the singles player, can still cause damage to your records. Even with the best engineering in the world, stacked records can damage the surfaces of each other. Warped records, or dust trapped between records cause scratches and other damage to the tracks. The result is a very much shortened life for the records.

Undesirability of stacking albums. The problem is much more prevalent with albums than with singles. At least the 45 rpm records have a raised centre which helps to prevent two recorded surfaces from coming into contact with each other. But no such raised lip exists on LPs. When you play one on top of the other, the top record is being gripped and turned by the tracks of the one underneath.

The final major disadvantage is the handling. Somehow, even with the best of intentions, you do not take as much care with eight or ten stacked records as you do with one. Every time you finger the tracks you leave grease marks. At best, these will shorten the life of your stylus; at worst, they will ruin your records.

Plug in a centre. It is possible to buy 'plug in' centres for turntables. This means that you could retain the short centre pole for your album playing, and the long, automatic changing pole for your teenage daughter's/sister's/brother's stints at singles playing.

Should you decide that this is the best compromise, you might also consider having two cartridges. One could be for your use (locked away safely when you are not actually playing records), and the other for singles.

There are other types of changers available, if you insist on having an automatic system, which do not stack records when playing. But, they are incredibly expensive and take up a lot of space. One type works on the juke box principle, laying one record at a time on the platter, holding the others you are waiting to hear on a special arm on one side of the plinth.

PART 3. THE PICK-UP

As we indicated earlier, many Hi-Fi enthusiasts feel that automatic decks are not really Hi-Fi, let alone automatic changers. These purists suggest that you should only consider a pick-up that is manually controlled. The argument is that an automatic arm causes unnecessary strain on the stylus, especially when it comes to the end of its run and lifts itself off the record to return to its rest.

Of course, if you want an automatic arm, you have no choice but to buy a turntable with an arm fitted. You will not be able to select the pick-up arm of your choice. However, do not think that there are no good turntable/ pick-up combinations on the market. Some makers have built an excellent reputation on them. It will no doubt be possible to find one which will please you in every way.

Certainly, if you do not want an automatic changing system, you may consider it pointless to have an automatic arm. So why not choose a manual arm to provide you with optimum quality for your money, just as you would any other unit in your Hi-Fi outfit? Let us take this opportunity to examine pick-up arms and see how it is possible to find exactly what you want.

THE ARM AND ITS FUNCTION

The pick-up arm has one basic fundamental requirement. It must be able to move freely about its axis to allow the stylus in the cartridge to 'track' faithfully the information in the record groove.

That may seem obvious to you, but there are certain problems which stand in the way of the arm achieving this function effectively.

To begin with, the arm has to travel in an arc, and therefore, the positioning is critical. The length of the average pick-up arm is 9 inches, and because it travels in an arc, it tends to get out of parallel at the end of its movement on the record.

You can see this for yourself. If you study the movement of the pick-up arm across the record you will see that the angle of the head in relation to the record groove is not the same at the end as it was at the start. When the record is finished, the angle of the arm to the groove is much more acute. It introduces a factor known as 'end of side distortion'.

DISTORTION COMPENSATION

All sorts of attempts have been made to compensate for this. Pick-up arms have been bent, heads have been set at an angle, but the problem has not been entirely 'licked'. The only real answer is a parallel tracking arm. This is an arm which moves along the record, parallel all the time to the area of the groove which it is touching.

While this has been achieved practically—at least one turntable has been fitted with such an arm—it still remains very much a dream for the vast majority of Hi-Fi enthusiasts.

THE INTRODUCTION OF BIAS

One major set-back in the attempts to adapt the 'arc moving' pick-up arm to something akin to the parallel movement is the introduction of an inward force, or bias. This occurs because the arm, when tracing the arc, overhangs the centre spindle. The bent arm exaggerates bias and, as a result, compensation devices have to be added.

So, these are some of the problems the manufacturers of pick-up arms have to contend with. The trouble is, this is only a part of the story. We have not even dealt with such things as mass, inertia, balance, tracking force, etc. as

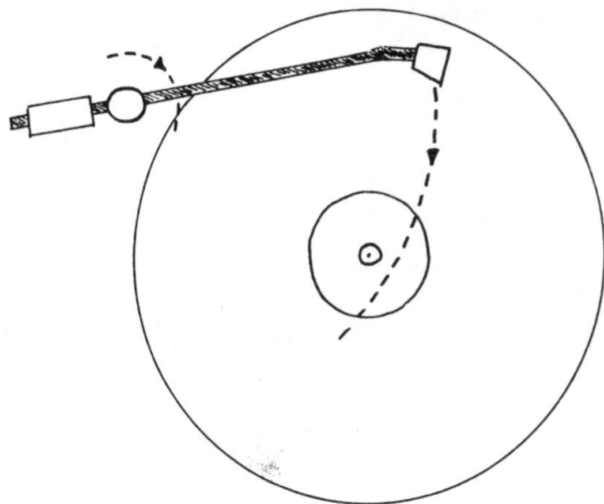

Fig. 21 Bias, the natural movement of a pick-up arm.

yet. Therefore, as a prelude to this, and to help us obtain a clearer picture of a pick-up arm, let us have a close look at its anatomy.

DISSECTING AN ARM

A pick-up arm has one basic job to do—support the stylus which tracks the record groove. It may or may not be sold with a turntable attached, although it will always be found on automatic turntables.

The arm consists of many parts. To start with, there is the arm itself. This should be sturdy, yet light in weight, and is usually about 9 inches in length.

Then there is the pivot which should allow the arm to move quite freely, within limits, laterally and vertically, with the minimum of friction. A counterweight, usually adjustable, makes the third part. It is fitted to the short end of the arm (on the opposite end to the cartridge) and acts as a fulcrum on the pivot to balance the arm in order to reduce the tracking weight or force.

At the other end, we find part number four—the head. This is the unit which is, in reality, two units in one. It is a shell (holder) and a cartridge. Inside the head is fitted a

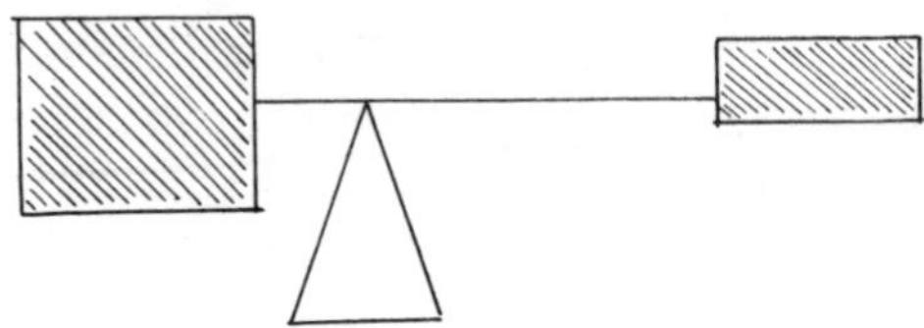

Fig. 22 The fulcrum principle.

device which converts the movements of the stylus in the groove to electrical energy. This has an added critical character in that it has to be matched to the input of the amplifier. But, more about that later.

The shell may or may not be a separate item. Sometimes they are fixed permanently to arms. While this may limit flexibility, it does have the advantage of reducing the weight the locking assembly has.

Added to these general parts, you will often find as a part of the pivot a bias compensator (also known as an anti-skating device), a stylus tracking force dial, and a special arm lifting lever.

INTO ARM SPECIFICATIONS

Right, knowing the arm as we do, let us quickly run down a set of pick-up arm specifications and see how we can learn something about the unit from them.

Friction—vertical:	6–8 mg
—horizontal:	5 mg
Effective mass:	18 g

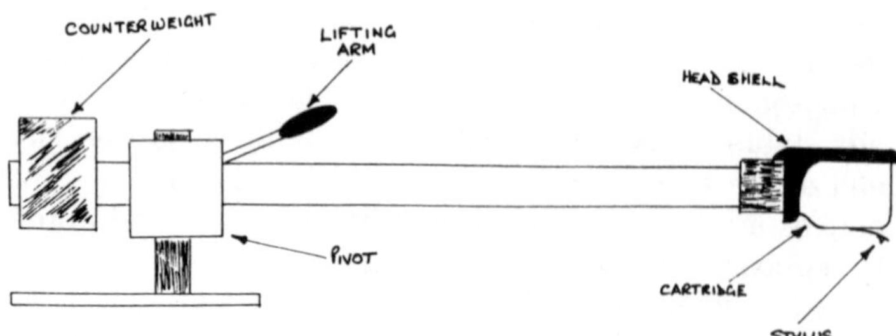

Fig. 23 The geography of a pick-up arm.

Effective length:	8.6 in.
Inertia:	9 × 41 gm·m²
Tracking force scale accuracy:	—0.1 g
Max. tracking error:	0.35°/in.

Friction. The effect of movement and resistance to movement in any direction. Various systems of bearings are used in pivots to keep the friction down, but it still exists. In our model specifications, the friction is low, which is significant of most modern pick-ups. However, it can vary from model to model within a range and, as a result, at times, recommended cartridges do not track as well as they should.

Friction, as you can see, is shown in milligrams, indicating the degree of resistance.

You can run a simple test for friction when you are looking at pick-up arms. Balance the arm off its rest so that it is 'floating' free. Then give it a slight push with one finger. The arm ought to move freely, vertically as well as horizontally without any jerks, or without coming to a sudden halt.

Effective mass: Effective mass is the total distribution of weight in the arm. It can be concentrated in the area of the pivot by adjusting the counterweight, thereby lightening the tracking weight.

Effective length. The distance between the pivot and the stylus.

Inertia. In a nutshell, it is the effective mass multiplied by the (effective length)^2. In practical terms, and to understand how it plays a part in pick-up arm specifications, we need to study it in relation to a warped record.

The cartridge will ride over the warp at anything up to 20 in. per second. As it rides up the warp, it has the tendency to be pushed upwards. The effort required to do this is determined by the inertia of the arm, and its effective length.

As the arm accelerates, it is counteracted by the inertia, resulting in an increased tracking force.

However, when the stylus reaches the peak of the warp, it should stop moving in an upward direction and stay in the groove of the record on the way down. But, the upward movement continues, and the cartridge jumps the tracks. Strain on the cartridge caused by this action can be reduced by keeping the effective mass and the inertia low.

In simple, non-technical terms then, inertia is demonstrated by the continued movement of the arm past the desired point caused by its weight or mass. The more mass distributed to the cartridge end, the more inertia you can expect.

Tracking force scale accuracy. Tracking force is the average weight exerted on the stylus by the pick-up head when tracking. This specification shows the discrepancy that exists between the calibrated tracking force on the adjustment dial, and the actual force.

Maximum tracking error. We are back to the arc path of the arm and the increasing angle from true parallel the stylus is to the record groove, per inch.

ON TO THE CARTRIDGE

The cartridge is the part which picks up the sound signals from the stylus and transmits them as electric energy to the amplifier.

There are a variety of pick-up cartridges to choose from. At the lower end of the scale is the simplest, and least expensive—the ceramic cartridge. It is not really a Hi-Fi cartridge. It has a high output—70–100 mV—and needs high impedance—some 2 MΩ.

At the next stop on the ladder we find the moving coil cartridge. This type is very popular and consists of two coils (stereo) attached via cantilevers to the stylus. Due to the stylus movement in the record groove, these coils fluctuate to the surrounding magnetic field created by a permanent magnet. The resultant voltage is then fed to the amplifier.

Finally, there is the magnetic cartridge. Probably the most popular with Hi-Fi enthusiasts, it is also the most expensive of the three. It tracks at very light weights—1 gram—and should be fitted to the better quality pick-up arms. It functions in a similar manner to the moving coil type.

THE BRAIN AT WORK

The pick-up cartridge is said to be the brain of the system. Any distortion, or coloration, introduced by the cartridge allows no hope of a cure by the rest of the system, no matter how good or expensive it might be.

When the stylus tracks the groove it bounces from side to side off the irregularities 'engraved' in the walls of the groove. These irregularities are the sounds you want to hear. The bouncing about happens hundreds of thousands of times during the period of play.

Obviously, you cannot hear the actual bouncing about—except in the form of music. The cartridge turns these vibrant bounces into electrical impulses which are boosted by the amplifier to a suitable voltage to drive the loudspeakers.

OUTLINING THE CARTRIDGE

Even with this item, you cannot escape specifications. It is a separate unit after all, and if you are going to select your own cartridge, then you will need information about the different types available to be able to narrow down the choice.

However, at this stage, specifications are beginning to become very familiar. Even those you have not seen before are no longer hard to work out.

Cartridge specifications are broken down into three parts—the description, the performance, and matching recommendations. They look like this:

Stylus configuration: .0002 in. × .0007 in. Elliptical
Solid nude diamond

Frequency response:	5 Hz to 20 kHz ± 2 dB
Tracking force range:	0.75 g to 1.5 g
Channel separation:	Nominally 30 dB at 1 kHz
Output voltage:	3.5 mV each channel at 5 cm/sec peak velocity.
Load recommendations:	47 kΩ

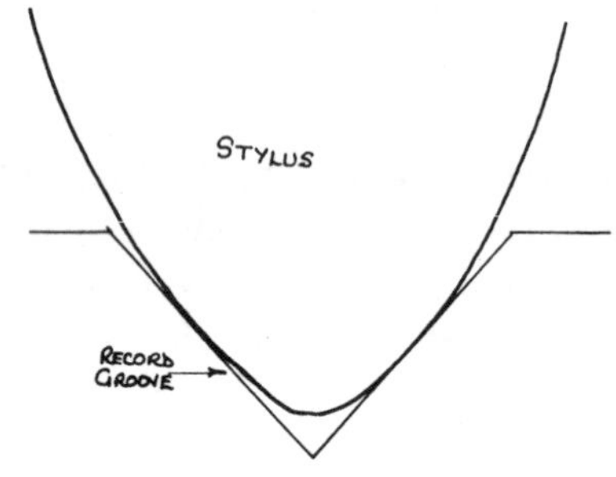

Fig. 24 Close-up of a stylus in a record groove.

Stylus configuration. This specification relates to the description of the stylus. As you can see, the tip dimensions in inches are stated first. The larger figure (.0007 in.) is the radius front to back, and the smaller (.0002 in.) is the radius side to side. The sides of the stylus actually trace the record groove and in order to be seated effectively, the stylus has to conform as closely to the groove shape as possible.

Styli have been through a variety of shapes starting with conical, or hemispherical. The term describes the shape of the tip where it comes into contact with the groove. But, this has largely been replaced now by the elliptical stylus which has a more oval configuration and sits in the groove in practically the same way as the original cutting stylus. As surface contact with the groove is improved in this way, it results in less wear on the record.

In recent years, an improved elliptical stylus has made its appearance. It is known as the 'Shibata' stylus and, while designed primarily for Quad systems, it is excellent for stereo. The profile of this stylus (and others of its type) is such that it gives an even better groove-surface contact.

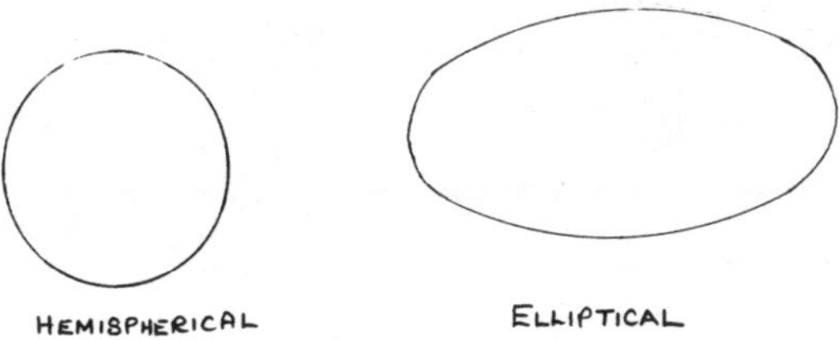

Fig. 25 Stylii—plan view.

The vertical points of contact are much closer to the record groove shape, while the lateral (front to back) contact has been kept to an absolute minimum.

This is the closest shape yet to the original cutting stylus. The problem is that manufacturers cannot duplicate an exact replica of the form since its sharp edge would immediately destroy the groove sound signal patterns.

A hard edge. Standard elliptical or Shibata styli are cut from diamonds—material that should be considered a must for Hi-Fi reproduction. Sapphire tips are available, but because of their comparatively short life they are generally considered to be unsuitable for Hi-Fi.

For the ultimate in quality sound, your stylus must be inspected regularly. Your dealer may be able to do this through a microscope specially designed for the job. The only alternative is to change the diamond stylus after every 500 albums, or their equivalent.

If you use a stylus in a worn or chipped condition, it will do untold damage to your records. The worst of it is, you cannot even hear it happening until it is too late. So, be warned.

Frequency response. This one is so familiar that it needs no explanation. You can refresh your memory by checking the explanation in Chapter Two.

Tracking force range. Measured in grams, this is the weight range at which the stylus will track effectively without causing damage either to the stylus or the record.

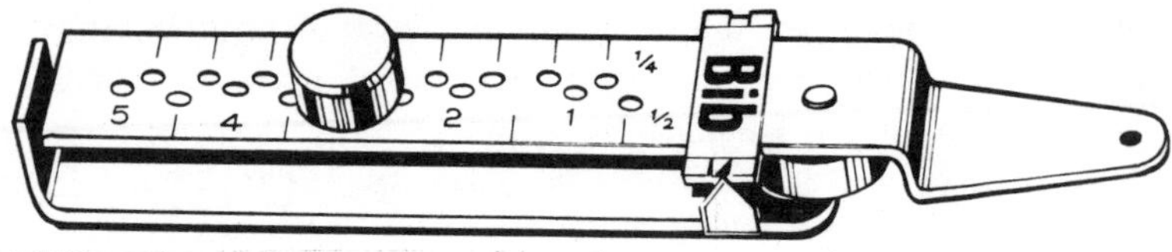

Fig. 26 Bib stylus balance.

Channel separation. It was in Chapter Three that you first came across this term, only then it was called 'Stereo Separation'. While names can change, with a little bit of thought, you will be able to decipher the meaning and see them for what they really are—old friends. Channel separation then, refers to the spillage of sound signals from one stereo channel into the other.

Output voltage. This, together with the next specification, 'Load Recommendations', indicates matching capabilities. Cross reference these two figures with the input recommendations relevant to the 'Cartridge' in the amplifier section of this book.

'Output' and 'Load' specifications of the cartridge are the vital factors in matching to the amplifier. It is not the turntable itself, nor is it the pick-up arm. You can virtually choose which you like of these two. But, once you have bought your amplifier, then you must find a cartridge with a stylus which is compatible, and which, of course, fits the arm you have chosen.

QUADRAPHONICS

No modern book on Hi-Fi would be complete without some remarks on Quadraphonics—that is, all-round sound. It is the up-and-coming thing and is a step onward from stereo sound. As the greatest problem in the advance of Quad sound research has been its application to records, it can rightfully be introduced here. However, you should remember that it can, and does, apply to tape.

The development of Hi-Fi sound has been fascinating. First came mono. Those were the days when you only needed one speaker. Then, in an attempt to capture the realism of musical sound, stereo was developed.

Stereo sound is basically produced by recording two tracks through two microphones at the same time. Of course, there are some mixers and other circuitry in between. However, in effect, the left microphone picks up the sounds emanating from the left, and the right microphone catches the sounds from the right.

These are transferred to records and tapes in the same way. On the right-hand side of the groove or tape track goes the right-hand microphone sounds, and on the left groove or track goes the left microphone sounds. These signals are then reproduced through the relevant 'channels' and are emitted as audible sounds through the respective right and left speakers.

Simple, isn't it?

Introducing Quad. But, for the Boffins, this was not enough. While stereo might contain the element of realism, it is all on one plane. The Boffins wanted depth as well and the answer was two times stereo, or quadraphony.

It started in a small way at first. A third speaker (or second pair) was linked up via the two live terminals on the primary speakers and positioned behind the listener. An impression of depth was the result with 'concert hall' ambience filling the room. This technique is called the 'Hafler effect'.

Then special circuitry was added and the ambience was improved. The listener could feel himself surrounded by music. Such experiments were not unnaturally given the name of 'Ambience Enhancing', and taking the basic principles involved, the circuits were improved yet again and the first of the two Quad systems made their appearance. They were the SQ (Stereo Quadraphonic) developed by the CBS Laboratories, and QS (Quadraphonic Stereo) developed by Sansui in Japan.

The matrix. Both use an intricate circuitry known as a matrix to decode the signals into four channels—one for

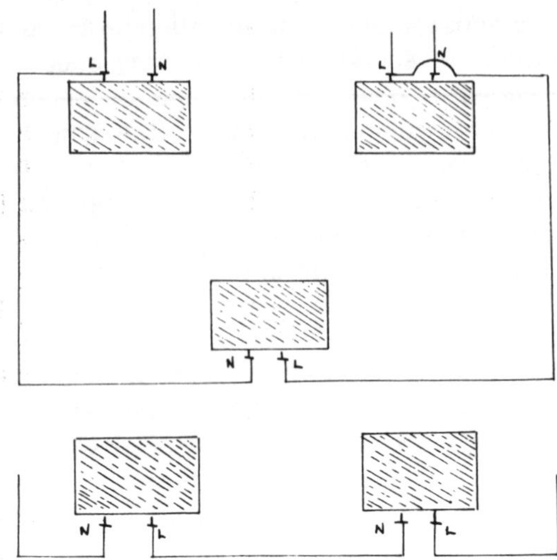

Fig. 27 Hafler effect, with one speaker and two.

each speaker. Records are now available which have been specially produced for these systems. Four channels have been encoded into two signals on the record or tape, ready for decoding by the matrix, back into four.

Two more systems soon made their appearance. They are the American-Japanese CD-4 (Compatible Discrete Four Channel) developed by the Japan Victor Company and RCA, and UD-4 (Universal Discrete Four Channel) by Nippon Columbia Co. These make use of the discrete signal method—that is four separate signals on four separate channels.

On open-reel tapes or 8 track cartridges (not cassettes) this is no real trouble. You simply need one track for each signal. But, when you try to apply the principle to the record groove, you are up against a whole new set of problems. A record groove only has two sides.

Four into two. This has been overcome by modulating the signals intended for left and right back speakers with a very high frequency, or supersonic wave (30 kHz). These two modulated waves are then 'added' to the regular signals and cut into the record groove.

If you play the record on a stereo system, only the regular signals will be picked up by the stylus. It cannot handle supersonic waves. However, a Quadraphonic set-up sorts them out and transmits them. A matrix is again employed to sort out the left and right front signals and the supersonic left and right back signals, and transmit them to the correct speakers.

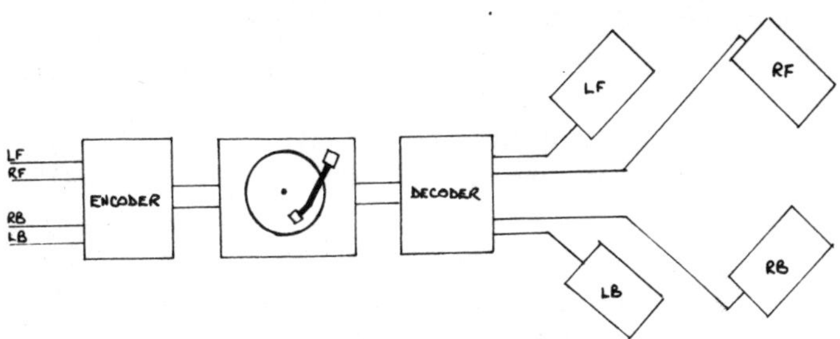

Fig. 28 Quadraphonic encoding and decoding.

Of course, you do need four speakers and ideally, they should be matched—especially to obtain the special effects on some pop records. But, it is possible to get away with two smaller speakers in the rear when ambience only is needed, since bass signals below 100 Hz are non-directional anyway.

CHOOSING YOUR SYSTEM

That, in a very small nutshell, is Quad. Of course, it is a great deal more complicated than this and you will 'hack your ankle' on such terms as 'Vari-matrix' (a control to increase channel signal separation); 'Logic Circuit' (sorts out to which channel a signal belongs); 'Logic Control' (improves signals and reduces cross talk); and so on.

But, that is the least of your problems. Your real difficulties will begin if you decide to 'go Quad' right from the start because, apart from all the technical points you have to watch out for, you also have to choose your system.

Make no mistake about it, you will have to choose. The four systems available are not interchangeable, or should we say compatible. All four systems have been developed independently and are currently fighting for supremacy. Unfortunately, this has held up acceptance by the public and one of the resultant major set-backs is that record companies are not about to make several different types of the same record to suit all Quad systems.

So, there has been a polarisation with certain major record companies backing their own systems and leaving the others to make up their own minds.

This means that perhaps your decision will be based on which system has the most records available (although you can play stereo records on all systems with an enhanced effect). At the time of writing, records for SQ are available in much greater numbers than the other three. You can, in fact, play QS records on an SQ system, and vice versa, but it is not ideal.

On the other hand, the availability of records may not be the criterion. Perhaps the discrete systems (CD-4 and UD-4) might give the channel separation and surround sound you are looking for. Again their records are interchangeable with each other, but not with SQ and QS.

But, perhaps you might decide to wait. After all, if you settle on one system and one of the others becomes the accepted one, you could be left with a 'lemon' on your hands. Even though a certain amount of interchangeability is possible, if you chose QS or SQ and discrete won the battle, you would have to start all over again.

Unfortunately, this is the precise reasoning which is holding Quadraphony back from being accepted generally. But, if you are rich enough to be adventurous, have a go. At least you will enjoy the depth of sound.

Adding Tape

Track system:	4-Track. 2 Channel. Stereo/Mono
Reel capacity:	Max. 7 in.
Tape speeds:	7½ ips. (19 cm/s), and 3¾ ips (14.5 cm/s) ± 1.5%
Wow and flutter:	Less than 0.12% rms at 7½ ips Less than 0.15% rms at 3¾ ips
Frequency response:	30 Hz to 23 kHz (±3 dB) at 7½ ips 30 Hz to 16 kHz (± 3 dB) at 3¾ ips
Signal to noise ratio:	Better than —56 dB
Erase ratio:	Better than —70 dB
Heads:	3 (Playback, Recording, Erase)
Output:	0.775 V
Load impedance:	More than 50 kΩ

INPUT

Microphones:	0.55 mV 30 kΩ
Line:	50 mV 200 kΩ

We have listed these specifications right at the beginning of this chapter to show you something. You can read them. You have progressed far enough that these figures mean something to you.

There are some things in this list which you have not encountered before. However, even then, what you have read in previous chapters has provided sufficient background for you to work out the answers without too much difficulty.

Read the specifications through again and make sure you understand them all. We will be departing from our established routine in this chapter and will not deal with tape recorders specification by specification. Such information will enter into the discussions, of course, but specifications will not be used as primary starting points for explanations.

THE RANGE OF RECORDERS

Just before we do leave them though, it is worth noting that this list refers to an open-reel tape recorder. There are other types on the market, as you are well aware, but whether they are cassette or cartridge recorders, with one or two modifications, the specification outlines are essentially the same.

We will be looking at the differences later in this chapter.

CLEARING UP SOME POINTS

One or two points in the list may need clearing up. 'Track System', 'Reel Capacity', and 'Tape Speeds' are clear enough, except you may be uncertain about the percentage figure in brackets after the latter figure. It refers to a possible fluctuation in speed. However, it is quite minimal.

'Wow and Flutter' should be very familiar. The problem, you will note, is more pronounced in the slower speed.

Frequency response is usually quoted in conjunction with a specific tape for recorders. It will alter slightly from tape to tape and is therefore of little real value unless you recognise the standard. It might be possible to find it quoted with your favourite tape to provide you with a comparison.

BE COMPATIBLE

The remainder are straight-forward, but do take careful note of the output specifications. Your tape deck must be compatible with your amplifier or you will not get quality sound. Too much signal will only give you distortion, and signals too low will give you a sound which is barely audible.

If you get compatibility right, then the remainder should cause you little difficulty. Therefore, we will press on with recorders in general.

FACING THE ONSLAUGHT

The first thing you will find when deciding to add tape to your system, is the plethora of machines available. You will find small ones and large ones, inexpensive ones and out-of-the-question ones.

Then there will be cassette recorders, cartridge decks, and open-reel recorders, originating from Denmark, Switzerland, Germany, U.S.A., England, and, of course, Japan.

We are told that they do certain things like sound-on-sound, instant playback, sel-sync, stereo record and playback, echo, mixing, etc., etc.

Where does it all end? The specifications were easy, and now this. The big question pops into your mind, "Where do I begin?"

The logical thing to do is to begin at the beginning. Start by asking yourself what you want the tape recorder for. Is it just for playing back music, or do you want it for recording too? If you want it for the latter, what source do you intend to use—radio? records? Or do you want to record live?

FUN WITH A RECORDER

With a microphone you can record the kids at a party, plays by the local drama group, music by the school orchestra, and a host of other things. Is this what you have in mind? Many Hi-Fi enthusiasts are solely interested in music, and completely ignore this aspect of tape recording. Make up your mind what you want to do, and start from there.

When you have the answers to these, ask yourself one final question. Do you want these recordings in stereo or mono?

EXPLORING RECORDERS

As with turntables, there are two basic types. You can either buy recorders with amplifiers, or without. In actual fact, it is not strictly true that a tape deck has no amplifier. It does have a pre-amp, which has sufficient strength to boost the signal enough to be handled by the main amplifier.

With the tape recorders which have integral amplifiers, you can carry them about and use them separately from Hi-Fi outfits. Yet, you can still plug them into other units if you want to.

A tape deck, while it is less expensive, is tied to your Hi-Fi outfit. It cannot be used remotely. However, if you have a fairly big open-reel job, you are not likely to want to carry it around anyway.

CASSETTES AND CARTRIDGES

If you have a cassette or cartridge deck, then the story might be different. The same rules apply as for open-reel recorders, but you are likely to want to carry a recorder of this type around with you—especially in the car.

Here you are open to a choice. You could buy an AC/DC portable recorder which will link to your Hi-Fi outfit, but still work in the car, in the street, or anywhere else you fancy using it. This sort of machine is a must if you are going to extend your recording activities beyond music.

The alternative is to buy two recorders. The first is a cassette deck which is plugged permanently into your outfit, and will not work separately. The second is a recorder made for the car.

If it is high fidelity sound you want from cassettes, especially in the home, then the second course is really for you. Cassette systems already have the disadvantage of miniaturisation, making good quality sound difficult to produce. You can weigh the odds in your favour by purchasing a machine which is tailor-made for Hi-Fi outfits. Compromises tend to compromise in all directions.

LOOK INSIDE

Well now, having confused you thoroughly with that, let us move on and lay bare the 'innards' of a tape recorder so we can see how it works and understand some of the physical electronic facilities offered.

The tape recorder has three main objectives—whether open reel, cassette, or cartridge. They are:

1. To take small electronic signals from a microphone, or other source, and amplify them to a level suitable to be recorded on to a length of magnetised tape.
2. To reproduce the signals now on the tape intelligibly via amplification (either built-in, or a separate unit) to a loudspeaker.
3. To have a mechanical transporting system which carries the tape across the record and/or playback heads.

The degree to which any or all of these are achieved successfully usually boils down to the quality of the machine. A tape recorder which is going to do all this to semi-professional standards is not going to be easy on the pocket. The inexpensive machine usually has some corners cut somewhere, and that is normally where it hurts—in the quality.

HEAD TO HEAD

But, whatever the quality, the principle of transferring the signal from source to tape is more or less the same. The magic lies in some relatively small objects which are generally called 'heads', but to the experts are known as 'transducers'.

In reality they are horse-shoe-shaped magnets wound with a very fine wire known as a 'coil'. The ends of the horse-shoes face one another forming an oval shape, rather like a miniature race track. The ends, however, are separated from each other and insulated by 'shims' or 'spacers' made from a very thin, non-magnetic material.

The average open-reel recorder has three heads—one for recording, one for playing the recording back, and one

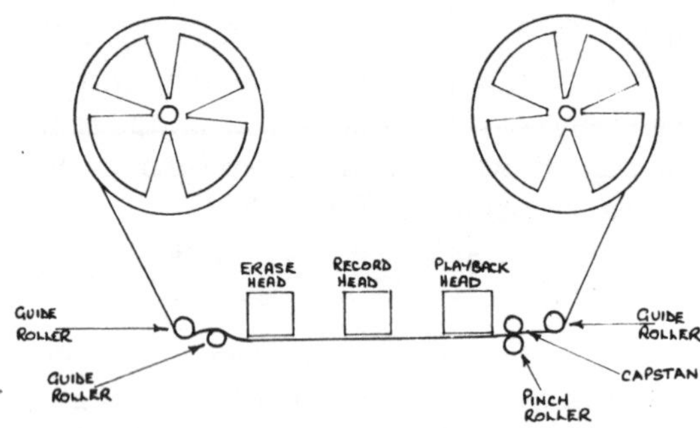

Fig. 29 Open-reel tape recorder head positions.

for erasing the tape. All are constructed in the same way, even though their functions are quite different.

THE RECORDING HEAD

When the audio signal is applied to the coil, it induces a magnetic field across the shims, or gaps. Then, as the tape is passed across the head, the magnetic field re-arranges the magnetic structure of the tape into a pattern which will remain indefinitely unless it is disturbed by another magnetic source.

Reading patterns. A head constructed in exactly the same way is used to 'read' the magnetic structures on the tape and reproduce audibly the information stored on it. This is, of course, the playback head.

However, it is not as simple as that. In the early days of experimental tape recording, the results obtained were far short of Hi-Fi. The information stored on the tape was distorted, unclear and fuzzy. Then, it was discovered that if a direct current was applied to the recording head in addition to the programme signals, the distortion could be greatly reduced.

This problem was not helped by the fact that the high frequency response was restricted to about 7 kHz. As the signal to noise ratio above this frequency was so ridiculously low as to be prohibitive, the overall signal to noise ratio left something to be desired.

The system used to surmount this difficulty is still employed in recorders today. A very high frequency—well above audio perception—alternating current is 'injected' into the recording head along with the intended signals. The results are a better response to higher frequencies and an extended signal to noise ratio.

A RECORDING BIAS

This high frequency alternating current, when applied to the recording head together with the programme content, is referred to as 'bias'. Some professional tape recorders have provision for altering this bias. This is because the amount of bias applied to the tape is really dependent on the tape itself.

In the less expensive domestic tape recorders, the bias is a fixed amount. It is adjusted to suit either the 'average' tape, or one particular type that the manufacturer chooses.

If this is the case, the recommended tape should be used for optimum results. The manufacturer will have inevitably set the machine for it. Nothing will be gained by using another brand, even though it might be advertised as having a lower noise level then the one specified. You might well gain a lower noise level, but the chances are you will be missing out on the responses to the high frequencies.

The importance of matching bias and the correct tape must not be underestimated. Professional studios consider a correct relationship so vital that they adjust the bias, not only from one brand to another, but also from batch to batch within the same tape brand. This is still standard routine despite the fact that tape manufacturing tolerances have improved enormously in recent years.

ELIMINATING SOUNDS

The high frequency alternating current is used in the erase head too. While it is possible to record over the top of the signals on a used tape (the magnetic structure of the tape is merely re-arranged), it is desirable to wipe it clean first. This is the purpose of the erase head.

Positioned in front of the recording head, the magnetic field created by the erase head arranges the poles of the magnetised particles in a north-south position. Any programme content thereon will immediately disappear.

TAPE TRANSPORTERS

You read earlier that one of the important things a tape recorder must do is to transport the tape across the heads at a constant speed. If the speed is anything less than constant, then you will experience our old friends 'wow and flutter'.

There is one sure-fire way of checking for 'wow and flutter', even if it is barely perceptible. Play a recording of a piece of piano music. If you cannot detect any noticeable raising or lowering of the pitch when single notes are played, or during soft passages, especially at the beginning or end of a full reel of tape, then the recorder should be alright.

But, if the slightest variation is there, find something else. It will not go away with use. In fact, as you use the recorder and the parts get worn, the fault becomes even more prominent.

WEIGHT BALANCE PROBLEM

In new models though, it is generally when the tape is at one or the other extreme points of its travel—all on one reel or the other—that this transport fault reveals itself. There is an uneven balance of weight, you see. First it is on the left motor, then it is on the right.

Moreover, the relative position of the tape to the pressure rollers and capstan is changed, and there is less back tension on the feed spool.

Your job is to find a recorder which can deal with this problem. After all, not everyone likes to hear the finish of the 1812 Overture in high 'C'.

INTERCHANGING TAPES

Another factor about speed is that it should be correct. It is possible that you will want to play one of your tapes on

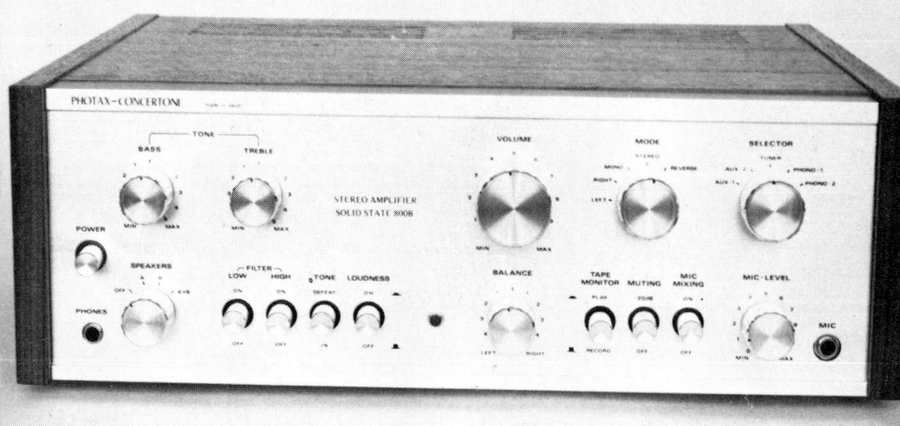

Photax-Concertone 800B stereo amplifier.

Alba UA 900 stereo amplifier.

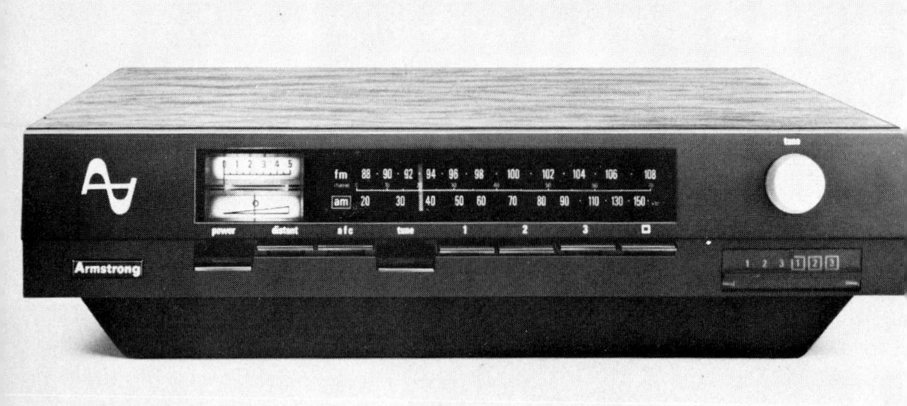

Armstrong 623 AM/FM tuner.

Country singer, Wayne Nutt, operates a mixing console at CBS recording studios.

‘Stampers’ ready to go into action at CBS record manufacturing plant.

Strathearn SMA 2 electric servo direct-drive turntable.

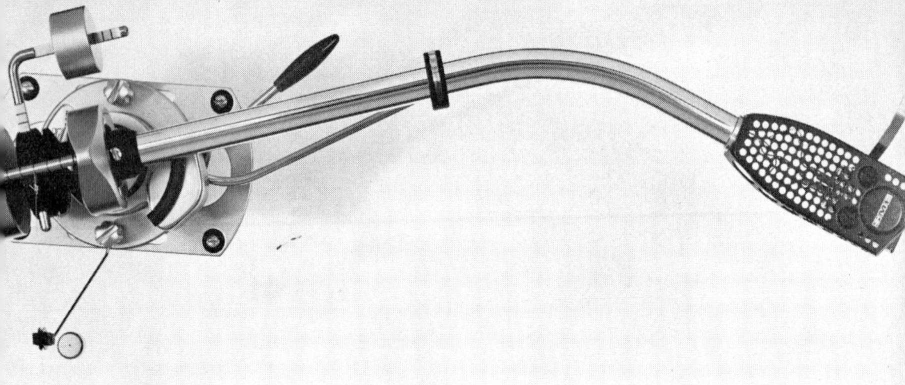

SME pick-up arm.

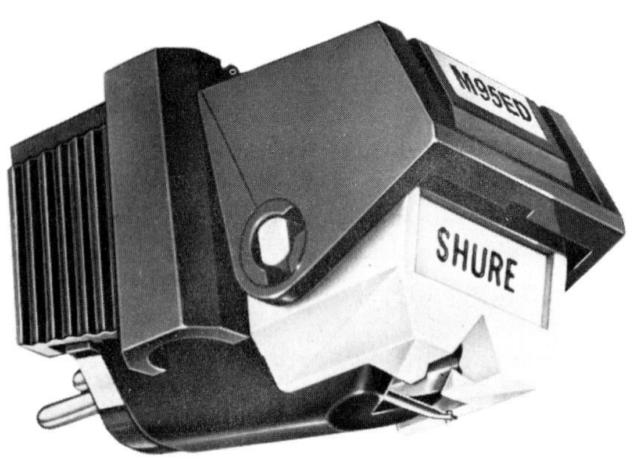

Shure M95ED cartridge.

Akai 1722L stereo open reel recorder.

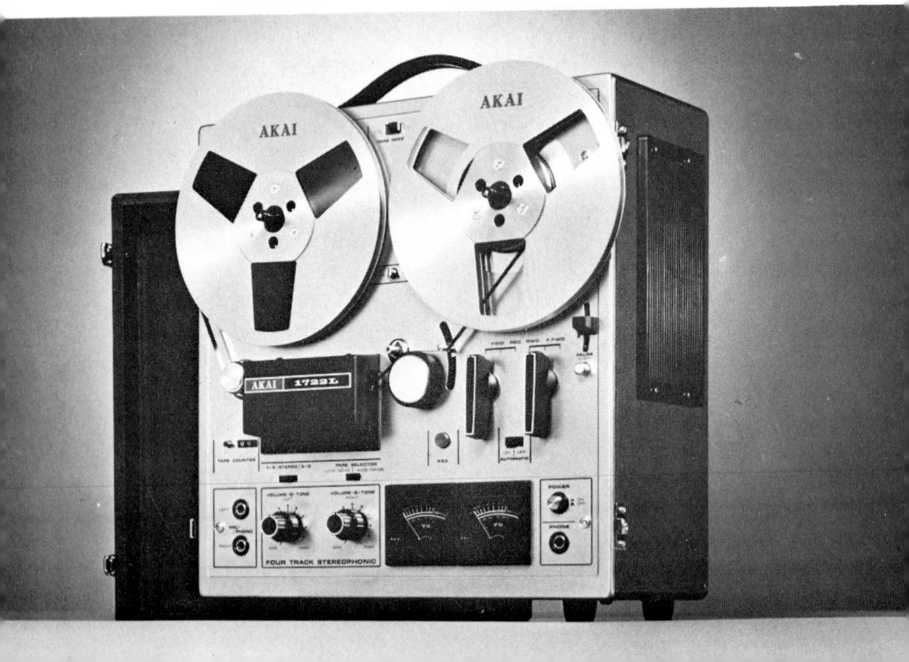

Yamaha TC 800GL stereo cassette recorder.

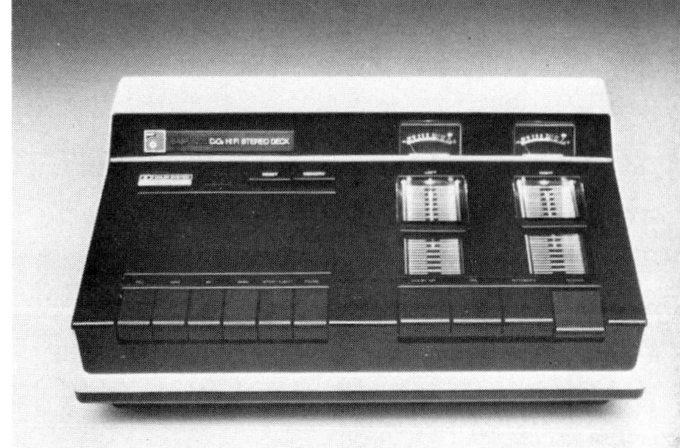

Basf 8200 stereo cassette deck.

Nakamichi TT 700 three-head cassette deck.

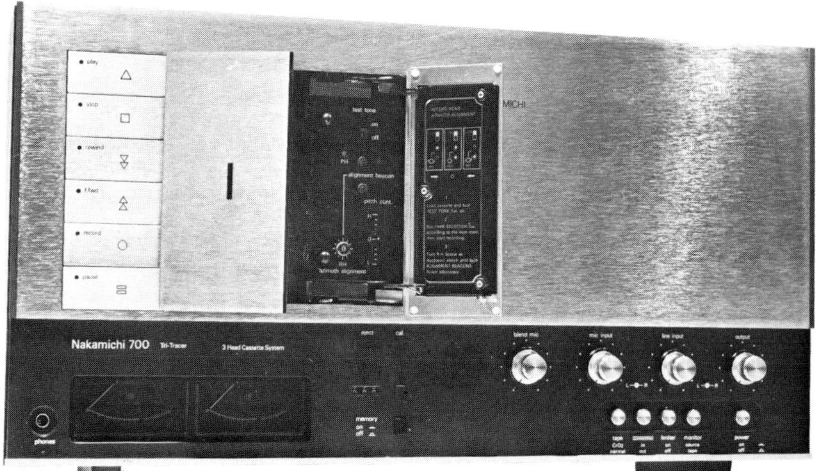

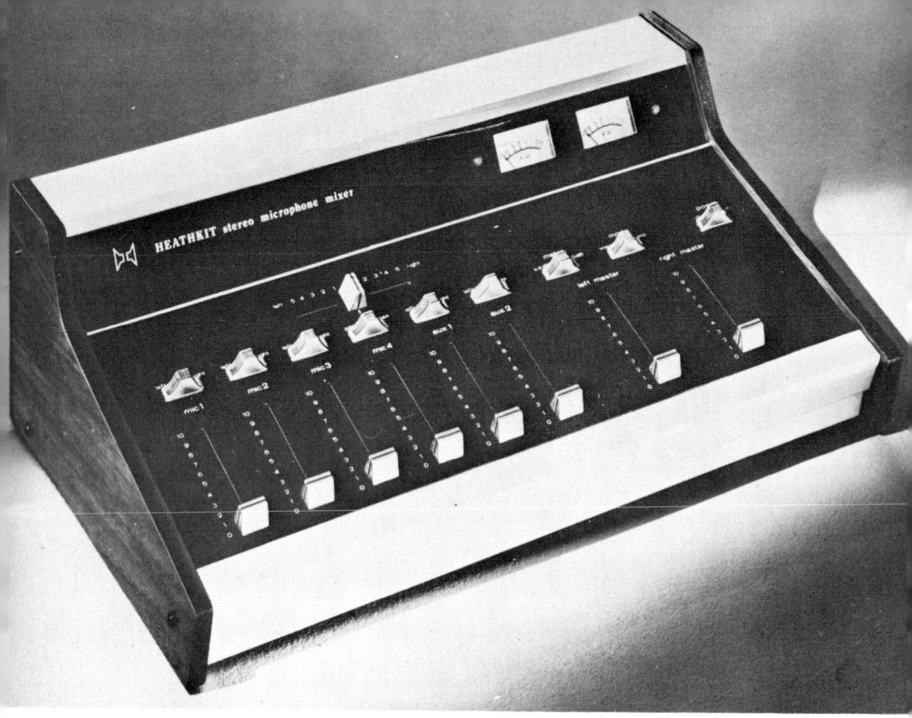

Heathkit TM-1626 stereo microphone mixer.

Celestion Ditton 11 (left), 33 (with grill removed) and 44 (centre) speakers.

Celestion UL6 speaker showing cone disposition.

Celestion Ditton 44 speaker showing cone disposition.

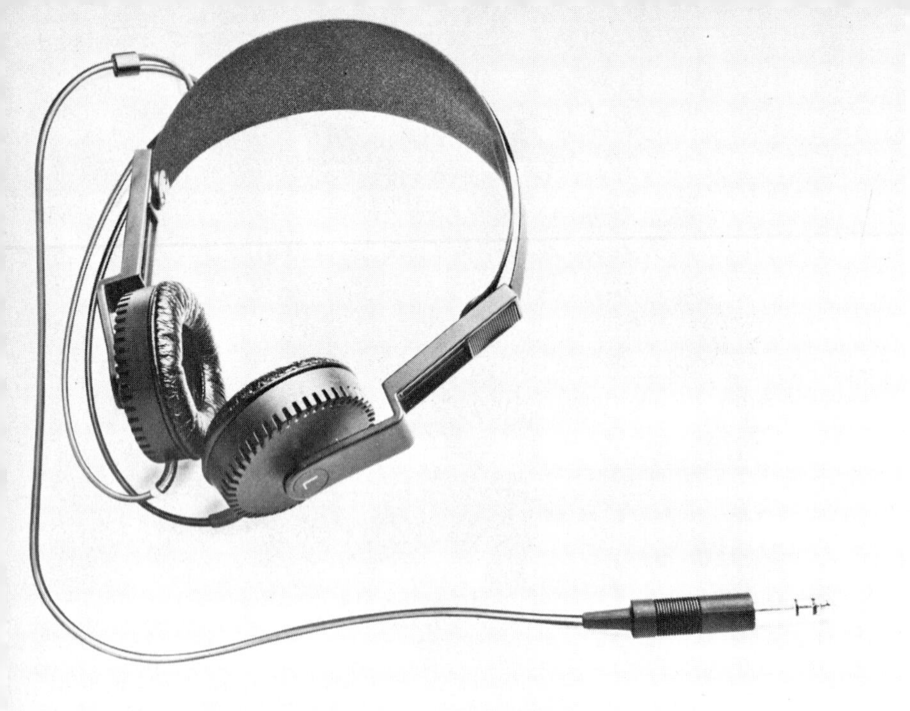

Yamaha HP1 headphones.

Groovac vacuum record cleaner.

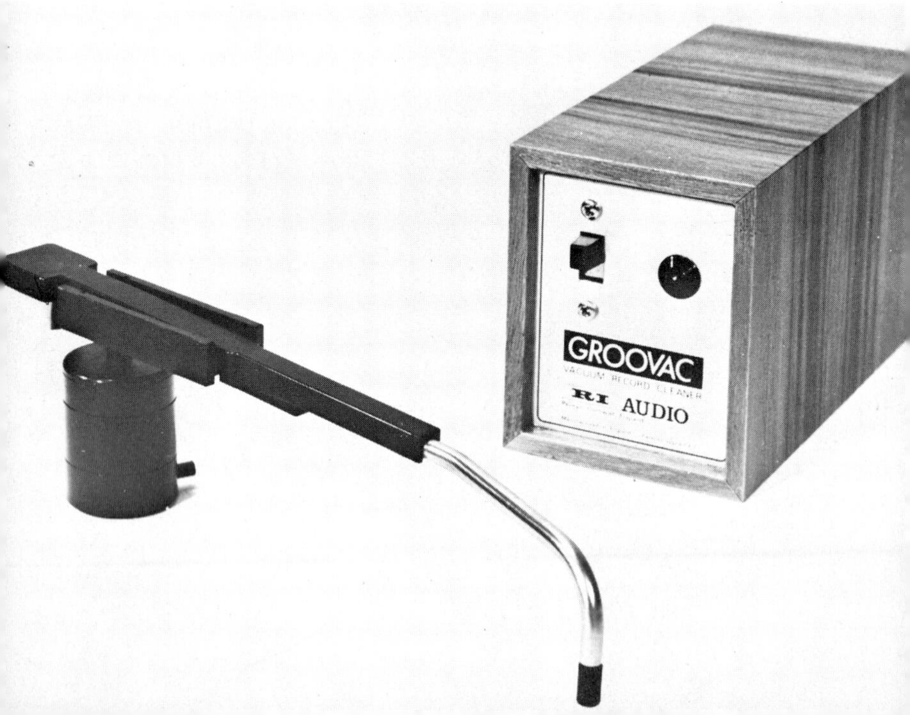

another machine—perhaps a friend's, or maybe you have bought another machine. Now, supposing your old one recorded at 9½ ips instead of 7½ ips. While your tapes were played back on the same machine, you could not hear the difference, because it was playing back at the same speed at which the tape was recorded.

But, play any one of those tapes back on a machine which had the correct 7½ ips speed and—well—it just would not be pleasant to hear.

An incorrect speed in a new tape recorder is unlikely, but just to be on the safe side, it would be prudent to check a pre-recorded tape from machine to machine when you are looking at different models to include in your Hi-Fi outfit.

CHOOSE THE FASTER

With the average open-reel tape recorder having two speeds, the temptation is to record everything on the slower one, thereby making the tape last longer. Try to resist the temptation if possible. It is a false economy. Yes, you will get more on the tape, but the quality will inevitably be lower.

To begin with, the noise level tends to be higher, the slower you play the tape. Secondly, you will have already seen from the list of specifications at the beginning of this chapter that the slower speed is more prone to speed fluctuations (i.e. the proclivity for wow and flutter is greater). Therefore, you would be wise to avoid using the 3¾ ips speed for anything which would spoil your listening enjoyment should this fault occur—this especially means music.

You can operate on the general 'rule of thumb' of putting all music on tape at 7½ ips, and the spoken word on at 3¾ ips. Naturally, there are exceptions to this rule. If you are a budding thespian, for instance, and your spoken word tapes consist of hard-earned complete collected live recordings of Chauncey Strongheart in all his Shakespeare parts, then the intonation of the voice will be important. In which case, it should be recorded at a speed of 7½ ips.

TAPE TRACKS

Before we move on to some of the technical aspects of recording, let us take a brief look at the geography of taped sound.

Most modern stereo tape machines record on four tracks. In effect, this means that there are four 'lines' of recorded signals running the length of the tape. Despite this, however, you will only be able to get two lots of recorded information on the tape.

The reason is because the tracks work in pairs. One track in each pair is for the right-hand stereo channel, and the other is for the left-hand channel. As both in the pair will have different signals recorded on them, you will get the stereo effect. So, the first time you run the tape through, you record on two channels. Then, you turn it over and run it through again. This time you use the remaining two tracks.

The tracks usually pair off alternatively. That is tracks 1 and 3 make one pair, and 2 and 4 another.

However, this only applies to open-reel recorders. Cassette recorders use a 1–2, 3–4 combination. In other words, the left- and right-hand stereo channels are on adjacent tracks. This makes cassette tapes suitable for playing on stereo or mono recorders and they are termed 'compatible'.

DOUBLE TRACKING

'Sound on sound' is another term you will encounter. It is merely an expression which describes a method of being able to record multiple sounds, one at a time, and have them all end up on the same track.

For example, let us suppose that you have decided to have a recording made of yourself singing in the bath. (It

Fig. 30 Stereo tape tracks.

Fig. 31 Stereo cassette tracks.

is advisable to have no water in the bath for safety's sake. Just make use of the echo.) Begin by making a recording on track 1 of the tape. Then, rewind it, and playback the original voice, re-recording it onto track 2 using the sound on sound facility.

Whilst this is happening, you can monitor the original on your headphones and record a duet with yourself. Now, you have two vocals on one track (track 2). The next step is to transfer the contents of track 2 back to track 1 while you are gaily singing and scrubbing (metaphorically) to produce yet another vocal harmony.

You could be a bathroom choir! Of course, people will talk—they will think you have taken this bath-sharing business to extremes. But, no matter. As long as you did not mix electric cables with bath water, you will live long enough to hear what you sound like a hundred-fold.

Well, perhaps this is exaggerating a little. There is a limit to the number of times you can do this, you see. Each time the voices are transferred from track to track, tape and electronic noises are amplified and join them on the new track. Therefore, only a limited number of these transfers are possible before the background noise and distortion becomes unacceptable.

Nevertheless, despite these limitations, it is a useful facility for building the individual sounds into an exciting 'orchestrated' whole. Incidently, it does not have to be confined to bath-time singing. You can use the technique to build up musical instruments played singly into a group sound, or even an orchestra.

BALANCING TRACKS

Sel-sync is a facility which allows you to record in the same way as sound on sound, but leaves you with only one vocal, or whatever, on each track.

You go through the same routine—that is making track 1, then listening to it while producing track two. However, each track will only have one sound on it. This means that you can make a much more accurate final balance. You can also add any effects to improve each individual sound such as echo, reverberations, etc.

When you have finished recording all your tracks, you can add them all together to produce the finished sound.

MONITORING

A facility which is most useful on recorders with three heads is a system of monitoring the signals just after they are recorded. Headphones are used, of course, and it is possible to check the quality of the recorded signal a few milliseconds after it is put on tape. The only alternative to this is to record the entire thing, then listen to it. If there is anything wrong then, you have to re-record the entire thing, or you have lost it altogether.

ECHO

If you are singing in the bath, you will not need it, but under other circumstances, you will find it useful to add a certain amount of echo to a recording. Well, it seems to be the done thing nowadays, so why not join in?

To put echo on a tape, you need, yet again, a recorder with three heads. What happens is, a part of the signal from track 1 is sent to track 2. Thanks to the gap between the record and playback head, the signal on track 2 will follow momentarily after that on track 1. The practical result is an echo.

MIXING SOUNDS

Some tape machines have mixing facilities like those you read about in the last chapter in the section on making records. It is a device which enables you to record sounds from two different sources at the same time. For instance, with this device you can sing into the microphone and use a musical backing from a record. The mixer takes both signals coming into the tape recorder and puts them onto a single track.

You may well ask why it cannot be done anyway, that is, without a mixing unit. After all, if you stand near the speakers and sing to the accompanying sound, both signals will enter the microphone.

Of course they will. But, will they be balanced? What will the quality be like? In answer to the second question, while the voice may be alright, the music will have a distant, tinny quality, and in answer to the first, balance will leave a lot to be desired. Try it and see.

The mixer takes care of all this. You will be able to control the balance easily, simply by operating a knob or lever. In addition to that, both sounds will be transmitted directly into the recorder making for a much better quality.

CASSETTES AND CARTRIDGES

So far, our discussions in this chapter have leaned very heavily towards the open-reel type of recorder. While the basic principles remain the same, there are sufficient differences in the systems to warrant a little time spent on exploring cassette and cartridge recorders.

These recorders have one distinct advantage over the open-reel type—they are compact and convenient. You merely drop in the cassette, or push in the cartridge, and press the play button. There is no threading, no chance of the tape coming free from its reel, no reel tension problems.

But, they do have one major set-back. They just do not come up to open-reel par in terms of quality. With miniaturisation comes compromises. Cassette tape speed, to allow reasonable programme lengths, had to be made incredibly slow—1⅞ips. This, as you have already learned, is not at all ideal for quality sound. Noise is increased as a result.

In addition, although frequency response in recorders is very much dependent on the tape, cassette and cartridge recorders performances in this respect are somewhat less inspiring than those from open-reel recorders.

THE DIFFICULTIES OF MINIATURISATION

The compactness of these recorders creates its own restrictions. For instance, they only have two heads—and one of those must be the erase head. This means that one head must have a dual function. It has a recording and a playback head within one screening can.

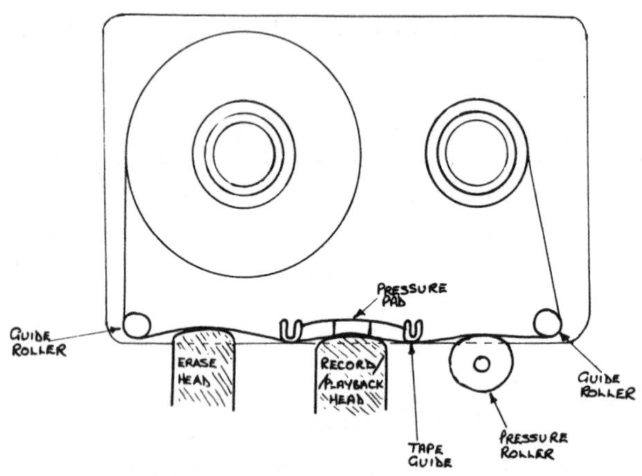

Fig. 32 Cassette recorder head positions.

However, cassette manufacturers are not to be outdone. Already a system has been devised where three heads can be built in. The playback head is the same size as and doubles as a tape guide. Needless to say, these three-headed machines are not cheap.

COMPACT MACHINES IN HI-FI

When considering cassette and cartridge recorders in terms of Hi-Fi, the cartridge machine comes a definite second. Research into convenience tape machines has almost exclusively moved in the direction of the cassette market. So much so in fact, that cartridge recorders are thought of as very much for the car (with a deck available for your Hi-Fi outfit to allow your cartridges to be played indoors), and the cassette recorder is becoming a Hi-Fi unit in its own right.

However, even though cassettes have lent themselves to the Hi-Fi market, and a number of manufacturers are now making good quality equipment, the battle for a sound fidelity as good as open-reel recorders still goes on. One of the avenues of research has been to reduce the inherent noise because of the slow speed.

The most common method used until recently was a system which reduced high frequencies. While this had the effect of a noise reducer, it did, in fact, modify the high frequency response. The problem is that when tape manufacturers are bending over backwards to improve the high frequency recording capabilities, such a restriction is counter-productive to say the least.

THE DOLBY SYSTEM

The latest system is the most effective. Known as the Dolby system, it has hitherto only been available on professional equipment in recording studios. However, it has now made its appearance on the popular market, mostly in cassette recorders.

Oddly enough, the Dolby system is not so widely used in open-reel decks. This is because the factors for noise reduction are already weighted in their favour—the three heads, the higher tape speeds, the wider tapes (¼-in. as opposed to ⅛-in. for cassettes) and recorded bands all help in this direction. Nevertheless some open-reel decks have incorporated the system, with effect.

HOW IT WORKS

It is a fact that whenever a recording is made, included on the tape, mixed in with the recorded signal, are background electronic noises. This is heard as a hissing sound. This can be heard during, and indeed can sometimes swamp, low passages of music.

When the Dolby system is installed, it takes and modifies the sound signal before and after it reaches the tape. As the sound signal is transmitted through the Dolby circuit for the first time, the lower sounds (softer passages) are boosted to a higher level than the noise. Then, on playback, these sounds as they pass through their second trip in the Dolby circuit, are reduced to their proper level, but this time the noise is reduced correspondingly so that it disappears from hearing.

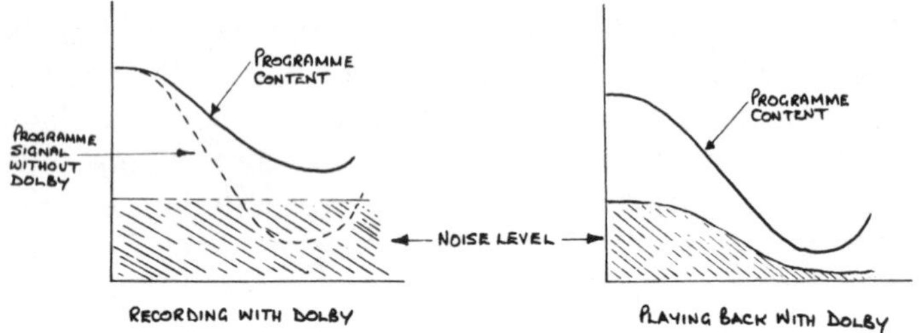

Fig. 33 The Dolby system at work.

The result is a cleaner sound, surprisingly clear of any form of background noise. You will have already heard the results of a Dolbyised sound. All modern records are produced in this way. Listen to one of your newest discs now. Can you hear a background hiss? If you can, the odds are it is not the record, but your equipment. In all probability, you will not hear it and you will realise what your recordings could be like.

TYPES OF MICROPHONES

The business end of any tape recording system is the microphone. It is considered so important that even the inexpensive tape machines can be fitted with very good models.

There are three basic types in general use today. They are the capacitor, moving coil and ribbon microphones. The most popular type is the moving coil, which also goes under the name of the dynamic microphone. It could be described as a loudspeaker in reverse. The programme signal vibrates a diaphragm causing it to move in and out between magnetic poles. As a result, a voltage is created which is sent to the amplifier in the recorder.

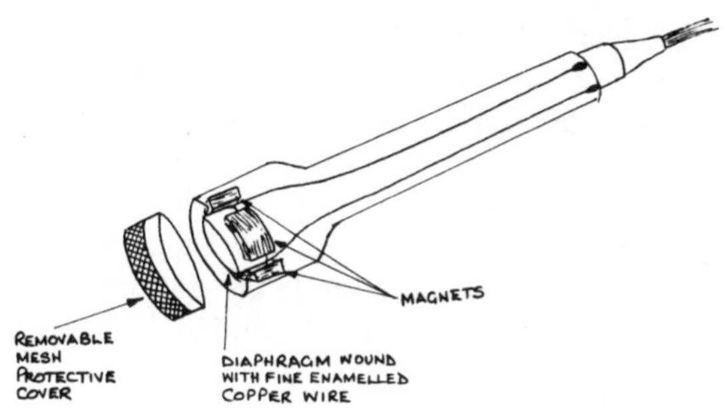

Fig. 34 Moving coil, or dynamic microphone.

The ribbon microphone works in a very similar fashion, but instead of a diaphragm, it has a small piece of metallic foil suspended between powerful magnets. From there on in, it is identical to the moving coil type. The resultant voltage fluctuations are transmitted to the recorder's amplifier.

Both these types of microphones produce very low voltage outputs. Therefore, they require plenty of amplification, which necessitates equipment with very low noise first stage amplifiers.

CAPACITOR MICROPHONES

The capacitor microphone used to be the exclusive domain of the sound recording studios. However, due to modern manufacturing methods, it has now been brought within the reach of the serious amateur.

This type of microphone has a diaphragm which is mounted on top of an electrically charged plate. When the diaphragm is subjected to air wave vibrations, the capacity (storage of electricity) of the capacitor is changed causing a current to flow. This current flow is minute and has to be amplified prior to its reaching the tape machine or mixer.

Usually, the tape recorder which is built for use with a capacitor microphone, cannot be used with any other. This is because of the necessity of this two-way flow of voltage. However, capacitor microphones can be used with any tape machine or mixer, provided it is accompanied by a special power pack which provides the correct voltage to the microphone to charge the capacitor.

HOW MICROPHONES CATCH THE SOUND

Not all microphones have the same field of sensitivity. Some can pick up sounds coming from any direction, while others will only 'find' the sound coming from directly in front of them.

Usually, the moving coil, or dynamic microphone, has a heart-shaped (cardoid) sensitivity field—known as a polar response. The ribbon microphone has a polar response shaped like a figure of eight. This makes it equally as sensitive from the front and the back, but not at all at the sides.

The polar response of a capacitor microphone can be tailored to suit the purpose for which it is to be used. On the more advanced models you can alter sensitivities to give you cardoid, figure of eight, or omni-directional response.

DIRECT RECORDINGS

Of course, when you have your tape or cassette deck linked into your Hi-Fi outfit, it is possible to record direct from your turntable, or tuner—or indeed, another tape deck. You do not have to use your microphone for this. Provision is made to plug an input lead (or leads) directly in to the recorder, the other end of which is plugged into the amplifier.

In point of fact, the same lead which channels the output signals from the recorder to the amplifier is used. It all goes into the same plug.

As you have learned before, you simply switch the amplifier knob or lever to indicate the unit you are playing and wish to record, then depress your record and play buttons on the tape deck. There is nothing else to it except to ensure that the input signal level is not so high that it goes into distortion, or so low that it will not record at sufficient volume.

Provision is made for checking this with a meter built in to your recorder. This flickers as the signals come into the recorder and a scale shows you when they become too much for the recorder to handle.

UNDERSTANDING TAPES

Open-reel tapes readily available for Hi-Fi outfits are usually ¼-in. wide and are either made of polyvinyl-chloride (pvc) or polyester with magnetic iron oxide particles 'suspended' on the sensitive (dull) surface. As the tape passes in front of the recording head, the particles arrange themselves in vertical patterns, the width of which is governed by the frequency of the sound signals.

The tape is uniform in thickness to prevent uneven reproduction quality, and iron oxide particles spread thinly and evenly over the surface achieve a sensitivity to low and high frequencies.

The base of the tape is both tough and flexible, capable of staying in one piece when it comes to a sudden stop, and resistant to curl and stretching.

Open-reel tapes are produced in four lengths—Standard (on 7 in. reels) at 1200 ft; Long Play at 1800 ft; Double Play at 2400 ft; and Triple Play at 3600 ft. While the reel size remains the same throughout, the tape thickness is altered to accommodate the greater lengths. Average thicknesses are (following the same order as above) 52 microns, 35 microns, 26 microns, and 18 microns.

Triple Play tapes are not really suitable for Hi-Fi because they are so thin. They are difficult to handle, a nightmare to edit, and they are also prone to a phenomenon called 'print through'. This is the name given to the transfer of magnetic influences from one wound layer to others and occurs when a tape is stored for any length of time.

The result is weak signals reproduced very faintly in subsequent layers. These can be heard just before or after the original, and go under the title of pre- or post-echo.

If you are a fan of triple play tapes and you have suffered from this problem, the 'echo' signals can be reduced by up to 5 dB by storing your recorded tapes inside-out. Then, when they are re-wound for play, reduction takes place.

Some modern tapes are manufactured with a matt backing. This has the advantage of spooling evenly when fast wound. The matt backing then prevents some edges from protruding from the main 'body' of the spooled tape, thus minimising the chances of damage.

DISSECTING CASSETTES

Cassette tapes are produced in much the same way as are open-reel tapes, and to the same standards and quality. The main exceptions being that they are about half the width (1/8-in.) and are housed in a plastic case.

The plastic casing has two small spools around which the tape is 'spooled' and 'unspooled'. The centres of the spools clip over two spindles which act as the tape transport. The tape moves from left to right from one spool, around a roller (sometimes there is a clip fitted to maintain tension), in front of a pressure pad, which is adjacent to the recording/playback head when the cassette is fitted in the recorder, around another roller, and finally onto the right-hand spool.

You can see the tape through a small window let in the top of the cassette to enable you to monitor its progress. They are available in four lengths—C 30 (15 minutes each side); C 60 (30 mins. each side); C 90 (45 mins. each side); and C 120 (1 hour each side). Added to these are a number of other intermediate lengths. You will find such cassette lengths as C 40 (20 mins. each side); C 90 + 6 (48 mins. each side); and so on.

However, from C 30 to C 120, the tapes themselves are of the thin variety.

CHROMIUM DIOXIDE

While all tapes have little to choose between them in terms of quality, and their sensitivity is relatively uniform, with noise and overload capacities before distortion being within 1 or 2 dB of each other, research has continued to improve on even this. Most standard tapes are produced in the 'low noise' category, but a relatively new type has been developed which improves sound reproduction still more. Known as the Chromium Dioxide tape ($Cr\ O_2$), it has a greater density of particles. Easily recognisable because of its almost black appearance (conventional tape is usually brown), it has an incredibly low noise level and is capable of reproducing very high quality recordings.

CARTRIDGE TAPES

We have already acknowledged the fact that cartridge recorders are not really in the Hi-Fi category, but a brief look at the cartridge itself will not go amiss.

Cartridges were constructed to be a playback system only and operate on the endless loop system. This means that you can play the entire content of a cartridge without having to turn it over. This is because it is made up of four sections of tape (two stereo tracks on each), and while this makes for convenience in the car, it can be extremely irritating when, as it often does, the tracks change awkwardly in mid programme.

The tape width is an improvement over the cassette—¼-in.—yet it still has all the convenience advantages of miniaturisation, and all the disadvantages in terms of quality (short cartridge life-span). However, it is transported at twice the speed—3¾ ips—which helps to improve the frequency response and to reduce the background noise. Unfortunately, on most machines, you cannot rewind a cartridge and it only has limited fast wind facilities.

EDITING TAPE

Any tape enthusiast will have to edit his tapes sometime. You may not want to make special recordings where tape editing is used in its strictest sense, but there may be the odd occasion when your tape or cassette tape will break.

To do it effectively, you will need a tape machine which has easily accessible heads and transport system. But, this does not let cassette systems out. You can still edit these tapes. It merely requires a little extra care, that is all.

The technique is to listen to the tape until you locate the area which you want to edit. Then, stop the tape and move it backwards and forwards manually until you find the spot you are looking for. Mark this position with a bright coloured (yellow or white) chinagraph pencil for easy identification, then cut it with a pair of non-magnetic scissors. Obviously, if your tape is merely broken and needs to be mended, you do not have to go through all this.

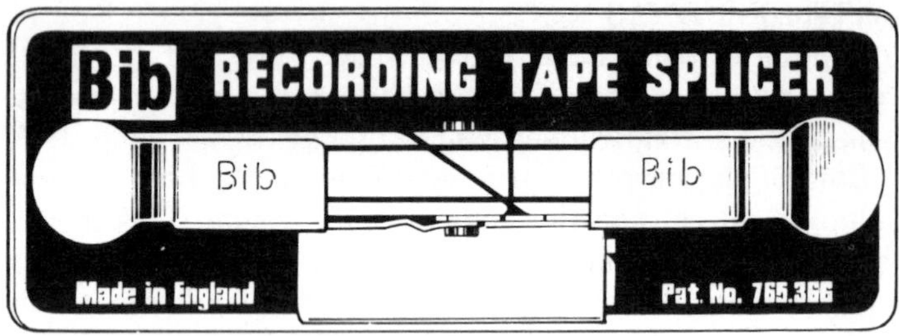

Fig. 35 Tape splicer.

Gather the two sections you need and position them on a jointing block (available from most Hi-Fi stores) and stick the two ends together with some jointing tape. The mark of a good tape editor is that you should not be able to hear the 'join' passing through the heads.

RECORDING SKILLS

Recording can be fun—but, it can also be a headache. Do not expect to pick up your brand new tape machine, plug in a microphone and get BBC-like results right away. It takes experience, practice and a certain amount of skill in employing recording techniques.

To help you to short circuit (if you will excuse the pun) the time-consuming learning skills, here are twelve direct routes to professional recordings.

1. If you want to record a speech with any sort of coherency, make sure you have a script. You cannot

expect to become a professional newscaster overnight (and, anyway, they work from scripts) and there is nothing worse than listening to a recording sprinkled with 'ums' and 'ahhs' as the disembodied voice fumbles through because he is not sure what to say next.

2. Should you want to make a recorded speech, then you might as well do it properly. Sit at a table with a script positioned almost upright in front of you. This prevents you from having to talk down at the table. If your wife has a go, remember that the female voice is generally quieter than the male's. We know it is hard to believe!

3. Some people tend to have a more explosive consonant sound than others. So, if you want to avoid having this magnified still more (and keep the microphone dry) obtain a foam rubber shield to cover the microphone.

4. A relatively 'dead' room is needed when you record speech. Your lounge will be best with its soft furnishings, curtains and carpets. However, beware of extraneous sounds. Microphones tend to pick up anything that is going—unless you have a very expensive directional microphone. Street noises just will not go down well with a carefully thought out and recorded speech. Find the quietest room.

5. Talking about extraneous noises, make sure your budding newscaster is well away from the recorder itself. You might even consider putting it in another room. Any recorder-operating noises will then be automatically isolated, and you will not have to be so dead quiet.

6. Provided your tape machine has separate recording and playback heads with electronic function switches, if your orator makes a mistake after he has been reading for some time, you can 'drop him in' just at the right place. You do this by winding back the tape a few feet and pre-select an appropriate pause, such as the end of a sentence or phrase. Then playback the tape, listening on your headphones, giving your speaker a visual cue when it reaches the pre-selected pause. As this happens, 'punch' the machine directly from playback to record, and carry on as before. With practice, you should be able to reach

the pause, cue the speaker and 'punch' in the record just as he begins talking. The results should be unnoticeable.

7. Forget all about trying to record the local pop group with your new tape deck and one microphone. You know from what you learned in Chapter Two how the professional studios undertake this task. They use a mountain of microphones, separation screens, etc. If you try to record the local pop group in your village hall, the sound will be hollow and boomy with the worst sufferers being the drums and bass, which turn up on the tape with a muddy, unclear sound.

8. Should you be determined to record groups, choirs, etc., then you should obtain at least four microphones and a mixer. Obviously, the better quality mixer you can get, the greater facilities it will have. However, you will have to pay for that quality. Use techniques for recording your groups, or whatever, as close to those described in Chapter Two as possible.

9. When you record a group of musicians who also sing, tape them playing their instruments only at first. Leave the vocals until later. You can use the 'sound on sound' facility to tape their voices. This time though, instead of accompanying themselves on their instruments, they can sing along with the recording they have already made by listening to it through headphones. By using this technique, they will be able to concentrate on getting the vocals just right, and you will end up with a tighter, much more professional recording.

10. Always keep the tape heads and path clean and demagnetised. Chapter Seven deals with caring for your equipment, so read the details and follow them. Dirty heads will mean that it will be impossible to get a good recording.

11. Never use old tapes for recording something that really matters. A bad splice, or a 'chewed' tape can ruin an otherwise good recording.

12. Regularly inspect your microphone and auxiliary equipment leads. A badly connected lead, or one that has had some rough treatment and might be hanging on by a thread

can break down at the most inconvenient time. In fact, the law of recording demands that it does!

Provided you treat it carefully, your tape recorder will last you a life time. You should get it inspected regularly, and any worn parts should be replaced. But, like everything else, if you do not buy one to do all that you want it to—and will want it to—you will find yourself paying out for another one because yours is now inadequate.

Nevertheless, make sure you use what you have paid out for. You will have wasted money if you buy an all-singing, all-dancing machine that does everything except make a cup of coffee only to play pre-recorded tapes on. Make the most of what you have and you will be surprised—you will have fun.

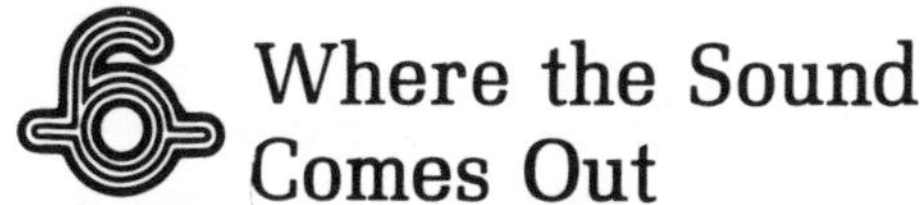

6 Where the Sound Comes Out

When dealing with a Hi-Fi outfit unit by unit, it is always difficult to put them in order of importance. Where do you start? No one unit will function by itself, so it has to be described in relation to another.

We began looking at the units by exploring the amplifier. This was described as the 'heart of the system'. In a sense, that is true, because without it we would not hear our sound-making units (turntable, tuner, or tape deck). In this case, we would be justified in saying that the amplifier is the most important unit and was rightly dealt with first.

Conversely, the amplifier is just a piece of electronic junk if it is not given a job to do. It needs a turntable, or a tuner, or a tape deck in order to function.

ENERGY ESCAPE ROUTE

But, and here is the rub, unless there is an 'escape route' and an ultimate converter for the energy coursing through these units—including the amplifier 'heart'—all of them are useless. They cannot be heard. A Hi-Fi outfit is dependent on those energy converters—the speakers—to be heard.

This makes speakers pretty important and if we were dealing with the units in order of merit, the loudspeakers should have been explored much earlier than this, perhaps even first. But 'order of merit' has not been the criterion, merely a logical sequence of 'heart', other 'internal organs', and now the 'external organs'.

VITAL ORGANS

We have said all this because we do not want the importance of the speakers minimised just because they appear last in our line-up of units. Make no mistake about it, they are vital, not just for obtaining sound, but for obtaining quality sound.

Your speaker can make or break your Hi-Fi results. Sound from the best system in the world will be lost with inadequate speakers. So, your choice is all-important if you want real high fidelity.

CRITERIA FOR CHOICE

As you search for a suitable Hi-Fi outfit, you have two choices open to you. You can buy the speakers first, basing your choice entirely on sound quality and efficiency, or you can buy them after you have purchased your amplifier. If you choose the latter course, then you have another choice-making factor to add—compatibility. The speakers will have to match the amplifier.

Given the latter choice because you have your amplifier already, the chances are good that you will find speakers which make the sound you like. This is because the choice is wide. There are more speakers on the market than any other single Hi-Fi unit. But, it may take you a little longer to find what you want.

Do not use this as an excuse to rush over making your choice though. Speakers are going to put the finishing touch to your sound quality and you should make sure they are just right.

VERSATILE SPEAKERS

You are going to be asking much of these units. They will have to deal with everything from a quiet passage of music played on a single instrument to a sudden burst from a full orchestra at full blast.

Sometimes you will want to reduce these sounds to a low enough level so that they merely fill in the background. At other times, you will want them so loud that they drown out all else. Your new speakers are going to have to cope with all your idiosyncrasies.

LENDING YOUR EAR

In a nutshell then, loudspeakers are very much a personal choice. Because every model presents some individual coloration of sound, this is one item in your outfit about which you cannot follow a friend's advice entirely. Do not feel put out by this. You can console yourself (and him) with the fact that even among professional sound engineers there are differences of opinion.

Therefore, if such experienced 'ears' disagree, there are also vast avenues open for us amateurs.

One basic problem with listening to loudspeakers in the dealer's shop is that they are unlikely to sound the same at home. Such things as the size and shape of your room and how it is furnished, will have a marked effect on the final sound.

TAKE THEM HOME

The ideal situation is to whittle the choice down to two or three possibles and, genuinely assuring the dealer that you fully intend to buy one of them, persuade him to arrange a home demonstration.

In all cases, whether listening in the shop or at home, you should use the same record as a test throughout. It should preferably be your own and contain some piano music. Any faults in the speakers, or indeed, the system, will soon be brought to the surface with the strains of a piano.

THE DRIVING FORCE

However, we still have to get to the 'listening point'. Before that stage is reached, there is much information with which you can arm yourself to help you to understand the job speakers have to do and to narrow down the field of choice.

The loudspeaker is the driving force behind the air which is needed to convey the sound coming from the record, tuner, or tape via the amplifier, to our ears. So, that 'electrical impulse' trail that we have been following in the last few chapters has finally reached its destination.

Now what happens?

MAKING WAVES

A basic loudspeaker unit is, in fact, made from specially treated and stiffened paper, or, in some cases, plastic. The centre of the cone is formed into a small cylinder, around the outside of which is wound a very fine enamelled copper wire. This is called a voice coil.

The ends of the voice coil are connected to two terminals mounted on the framework of the loudspeaker. It, together with the cylinder, moves to and fro freely between a magnetic field created by a permanent magnet which completely surrounds it. Electrical impulses (sound signals) fed into the coil from the amplifier cause the whole cylinder, and therefore, the cone, to vibrate in response to these signals. This makes the air in front of and behind the cone move in unison in the form of waves. The result? Audible sound.

MOVING COILS AND ELECTROSTATICS

This type of unit understandably has the name of 'Moving Coil' speaker. But there is another type which is becoming very popular. It is called the 'Electrostatic' speaker. The popularity of this speaker is not hard to understand since its performance is first class, producing an excellent and realistic frequency range.

The principle of the electrostatic speaker differs slightly from the moving coil type. Signals received from the amplifier are sent to a transformer which has a polarising voltage applied between it and the flexible plastic diaphragm. This diaphragm is coated with a conductive material and responds to the signal pulses via the polarising voltage.

CONE SIZE AND SOUND

A single loudspeaker cone can, if necessary, handle the entire audible frequency range. To a certain extent though, sound quality is dependent on the size of the cone. The larger it is, the better it tends to sound. This is because the larger cone is able to reproduce the lower frequencies more faithfully, and is said to have a good bass response.

Miniature transistor radio manufacturers realised this in the early days, and made available an optional extra called a 'sound booster'. This was no more than a bigger and better loudspeaker which improved the sound solely because of its size.

Some of today's Hi-Fi speakers (even relatively small ones in special cases) have an amazing bass response, but that is generally a combination of the size and/or quality of the cone, and the correct matching of the cabinet to the cone. During construction, they are tailored specifically for each other.

FROM WOOFERS TO TWEETERS

However, Hi-Fi speakers do not rely on one all-purpose cone to do the job. Quality sound requires a number of cones within one speaker unit, each handling a different frequency range.

Most modern speakers have three cones—a large, bass frequency unit, a smaller, mid-range cone, and a tiny, high frequency responsive cone. They are affectionately known as a Woofer, a Squawker, and a Tweeter respectively, and are designed to do their jobs most efficiently within their own, individual restricted ranges.

CROSSING THE FREQUENCY RANGES

Another important element in a loudspeaker is the Cross-over Network. This is a special circuit which controls the input frequencies and directs them to the correct speaker cones. There are cut-off points where a cone's reaction to a sound signal at a certain frequency diminishes in efficiency, and that signal must therefore be passed on to another cone which can handle it with no trouble.

POSITIONING THE CONES

The disposition of the cones within a speaker unit is important. Normally, as we have already said, there are three cones and these are disposed either one above the other, or in a pattern which makes maximum use of the available space to produce the optimum sound.

However, the latest cone distribution system to make its appearance goes under the name of the 'dual concentric' loudspeaker. It consists of a tweeter actually mounted onto a woofer cone. These two handle the entire frequency range.

The basic advantage is that the frequency range is transmitted from one source, thus eliminating the slightly disjointed effect possible with three separate drive units.

TRANSMISSION DELAY

One of the difficulties with speaker cones is getting the signal out as soon as it arrives. Unfortunately, some speaker cones tend to store 'energy' and there is a delay in the signal being emitted audibly. It is what is known as 'Phase' Response.

What is even worse, and therefore more difficult to control, the 'energy storing' time is not constant. It varies according to the frequency.

Now, this can create serious problems at the crossover points between speaker cones. Fortunately attention is being given to this phenomenon now, and it can already be corrected to a certain extent electrically in the crossover circuits. But, it still exists, and when choosing your speaker you should listen carefully after reading the details in the specifications about phase response.

All you can do is find speaker units in which the problem is so minimised as to be imperceptible.

CABINETS AND SOUND QUALITY

When considering speakers in general terms, the cabinets, or enclosures, must be brought into the picture. It may seem strange that a wooden cabinet will have a part to play in the final sound quality, but it is a fact. In truth, they are very important.

Many enthusiasts have purchased good quality loudspeaker units and put them into home-made cabinets. They have been pleased with the results, but this has probably been partly psychological because of a sense of achievement.

Then they hear a friend's system which has been constructed from the same units, but in cabinets designed by the manufacturer. The disillusionment is instantaneous. The sound is much better, and rightly so. The cones and the cabinets have been made to match.

DO DO-IT-YOURSELF

There is nothing wrong with making your own cabinets. Many enthusiasts do, very successfully. But, you must adhere to the manufacturers' specifications. It is no good making them an inch or two smaller than specified so they will fit between two shelves, for instance, if you expect the unit to reproduce sound at its best.

The reason is because the amount of air pressure on the rear of the cone is as important as the pressure in front. In 'enclosure' design, an inch larger or smaller than the optimum will make a difference to the sound.

Therefore, if you are handy with a set of woodworking tools, and you want to make your own, turn to outfits which are collected together and sold by manufacturers who have a good reputation. They will advise you every step of the way, even down to telling you what type of special filling to use for the inside cabinet acoustic treatment.

Do not think that you can do it all by yourself. Remember that the manufacturer has carried out experiments on a very large scale to find the right-shaped cabinet for their speakers. So it is hardly likely that the do-it-yourself tyro will improve matters by making the cabinets a different size or shape.

THE BASS REFLEX CABINET

Whether bought or home constructed, there are two basic types of loudspeaker cabinets. The first is the 'Bass Reflex' cabinet. This allows an extended bass frequency response on the pressure exerted from the rearward movement of the cone, which is then directed through a port in the front of the cabinet. The sound signal is pushed out to join the others coming from the front of the cone, reinforcing and enhancing the bass response.

THE INFINITE BAFFLE CABINET

The second type of cabinet labours under the title of 'Infinite Baffle', which means, total enclosure. The speakers are mounted on one side of a completely sealed enclosure which has the effect of keeping separate those tones emanating from the rear of the cone and those emanating from the front of the cone. Those 'rearward' tones are absorbed by the totally enclosed cabinet, thus avoiding any cancellation effect should the twain meet.

While the Hi-Fi world is full of opinion about which of these cabinet types is best, no one is prepared to lay his life on the line and issue a dying testimony about it. The truth is, there is not much to choose. It is all a question of taste. Listen to them both and fall into the ranks of those with whom you agree. The final decision as to whether you are a bass reflex or an infinite baffle fan must be yours—or rather, your ear's.

POSITIONING YOUR SPEAKER UNITS

A very nice way to hide your speaker units is to build them into your existing room decor. This has the advantage of putting them out of sight and, for that matter, away from little inquisitive fingers.

However, it also has its disadvantages. Once the speakers are positioned, you cannot easily move the furniture around to change the appearance of the room (and you know how often the wife wants that done!). So, if your wife wins the battle and re-arranges the room, bang goes your careful and time-consuming positioning.

THE SPEAKER AND THE FURNITURE

Whichever method you decide to employ—hidden, or in full view—placing the speakers is more than just a question of hearing the full magic of stereo, it also decides something about the quality of the sound you will hear.

Sound is reflected from the walls, floor and ceiling, in fact, sound reflection potential is made a part of the design of loudspeakers.

This is one of the main reasons why it is possible to detect the difference in reproduction from the same loudspeaker in rooms other than that in which it is normally used. Indeed, it is possible to hear the difference in sound when the speakers are moved to another position in the same room! The position of the furniture in relation to the speakers greatly affects the final sound.

For instance, with a large, full-range speaker system standing on the floor in the corners of a heavily carpeted room with floor to ceiling curtains and plenty of soft furniture, the sound reproduced will be slightly 'boomy' and lifeless. The reason for this is because the heavy furnishings tend to absorb the upper end of the frequency range. This, in turn, will cause the bottom end of the scale (the bass) to be over-pronounced.

THE SPEAKER AND THE WALL

Then, the fact that the speaker is backed against the wall plays its part in modifying the sound too. This tends to act as an extension baffle for the cabinet, giving the effect of even more bass.

Now, as the speakers are in the corner of the room, they have not just one wall to contend with, but two. This increases the bass phenomenon and will end up giving you an effect not a great deal different to sensurround. You will shake the foundations of the house!

TAKE ADVANTAGE

Actually, you can use this position to your advantage. This highly increased bass response will be effective if your system is bass light. Placing your speakers in a corner at floor or ceiling level will restore the bottom end of the scale enough to redress the balance.

As the majority of furnishings are on a low plane, you can improve the high frequency end of your speakers by raising them above that level. Suspend them high on the wall and the frequency scale will be improved.

Despite all this, a general 'rule of the thumb' for speaker positioning is, the moving coil type can be used without too much detriment in the corners and against walls, but the electrostatic type should be at least 2ft from either. The effects just explained are much more pronounced in electrostatic speakers and much more detrimental to the ultimate sound quality.

PLACING FOR STEREO

Distance is important if you want to achieve optimum stereo results. As a guide, the speaker cabinets should be placed 9ft to 12ft apart (given an average-sized room) and

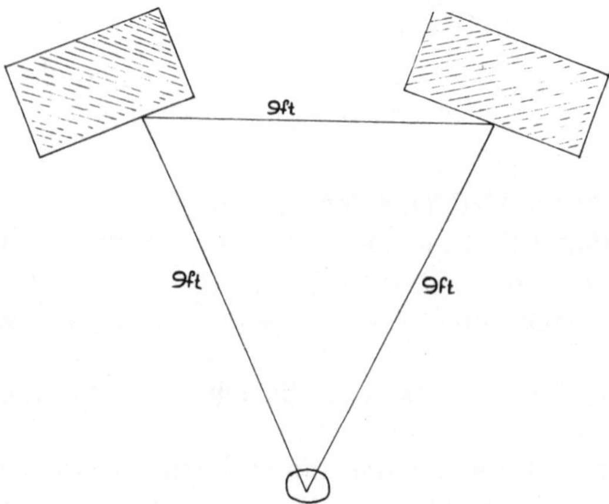

Fig. 36 Speaker positioning for stereo listening.

faced inwards at an angle to provide a sound apex at the same distance in front. In other words, a perfect triangle.

If you have to lay your speakers on their sides—say, on a bookshelf—make sure that they are positioned in such a way that their high frequency cones are nearest to each other. This allows a better sound disposition and, therefore, better stereo, because the high frequencies tend to have a narrow dispersion angle.

SPECIFICATIONS AGAIN

It is at this point in previous sections (with the exception of the last chapter) that we have turned to specifications to see what can be learned about the unit being explored. Despite the fact that as you read them, they will appear to be all too familiar, they provide an important method of helping you to understand and choose your speakers. So, we will again, follow the 'old routine'.

Power handling:	30 watts rms
Impedance:	8Ω
Frequency response:	30 Hz—20 kHz ± 2 dB
Sensitivity:	0.9 watts for 90 dB 1 metre/1 kHz on axis

These specifications can be divided into two basic categories. 'Power Handling' and 'Impedance' deal with matching speakers to each other and to the amplifier. The others deal with speaker performance.

So, let us begin with this business of matching.

COMPATABILITY

Power handling. In simple terms, power handling describes the maximum input figure. There are, in fact, two ways of describing power handling. First, there are 'watts rms', with which you are already familiar. Secondly, you will see it written as 'watts DIN'. The problem is that the two figures are not interchangeable, they only approximate each other.

DIN is a European measurement term which is gaining popularity here. It works on the basis that most people do not feed single frequencies into their speakers—which is how the rms figure is derived. Therefore, the resultant figure is not over-helpful.

Noise covering a reasonably wide frequency spectrum is what is used to establish the DIN measurement. This is fed into the speaker and the maximum signal it can handle without change over an extended period is the DIN measurement. However, a speaker given a DIN rating will be able to handle very much higher peak inputs for short periods.

Matching the figures. In point of fact, there is no need to match amplifier and speaker exactly. There is always a range within a certain amount of apparent incompatibility where it is possible. For instance, if your amplifier has an output of 30 watts rms, you can match it with speakers of anything from 20 watts DIN upwards.

Of course, you should not run this slightly mismatched system on full output for extended periods, or damage to the speakers is a certainty. But, as this is unlikely, the set-up will work quite adequately.

In keeping with this, some manufacturers are quoting a driving power range instead of, or in addition to, the power handling figure. Therefore, you will see such specifications as 15–50 watts rms, indicating that the speakers can be used with any amplifier which has an output figure within that range. Anything under would reduce the output signal below a realistic audible level, anything above might damage the cones.

Impedance. This can be thought of as a complicated amalgamation of anything in the circuits which offers any form of resistance to the signal coming from the amplifier. It includes resistance, inductance, conductance, capacitance, and so on. This conglomeration of resistance is termed as the 'load', in this case, offered to the amplifier, and is measured in ohms.

Most modern speakers are categorised as 4, 6, or 8 Ω, the latter two now taking precedence over the former. Other 'impedance' models do exist, but even then, the figure given is notional. The impedance is measure on the basis of a single frequency, and it does alter at different frequencies.

Therefore matching with amplifiers is not so critical as it first appears. However, it should be a lower stated impedance value in the amplifier 'looking into' a higher loudspeaker impedance figure—not the other way around. An 8 Ω amplifier looking into a 4 Ω speaker will cause higher distortion.

PERFORMANCE SPECIFICATIONS

Frequency response. There is nothing to say about frequency response that you have not read before. What you are looking for is how the speakers handle the frequencies, particularly at the extremes, such as organ notes down among the 40 Hz waves, and the high frequency violin strings.

Watch out too, for the decibel fluctuations in relation to the frequency response. If it is too marked, the lower and higher notes may be carried to frequency levels out of your hearing capabilities. It is no good having a frequency response of 30 Hz to 20 kHz if you cannot hear the notes at the extremes because a 6 dB fluctuation has literally made them almost disappear.

Sensitivity. This is also known as efficiency, and relates to the amount of power required to make the speaker reach a specific sound pressure level. The figure in our example tells us that for an input of 0.9 watts, an air pressure, which is converted into decibels, is obtained at a distance of 1 metre away from the speaker, on axis (directly in front), with an input of 1 kHz.

Some standards you can use for comparisons are that reasonably efficient speakers need something like 1–2 watts for 90 dB, a very efficient one needs 0.8–1.5 watts, and a relatively poor efficiency would give 2–3 watts.

Efficiency is also expressed as a percentage. Sometimes these percentage figures are so small as to be less than 1%. For instance, our 0.9 watts gives 90 dB and turns up an efficiency figure of 0.60%. This is perhaps the easiest figure to use for comparisons since you will be able to tell at a glance that a speaker with a sensitivity of some 0.7% is more efficient than one sporting 0.3%.

HEADPHONES

Headphones are fast becoming standard kit for Hi-Fi enthusiasts. No doubt, you will want to join them. Often, you will want to bask in the pleasures of music while the rest of the family are watching television. Unless you are one of the fortunate few with a den of your own, you are going to have to think of something else.

The answer is a set of headphones. Just plug them in and you are the only one who can hear the music. On most Hi-Fi outfits, when the 'phones are connected, the speakers are bypassed, reducing the signal to a level only high enough to be a reminder that the set is on should you put your headphones down and forget. Admittedly, you will be cut off from the rest of the family, but if they are all watching 'Coronation Street', or something, you will be effectively cut off anyway!

BABY SPEAKERS

Headphones are, in reality, tiny speakers. They are constructed in exactly the same way as a speaker, only in miniature. The most popular type is the moving coil, 'dynamic' headphones. They are less expensive than their electrostatic brothers, and are available in a wide range of prices and qualities.

Electrostatic headphones, on the other hand, have the advantage of being lighter in weight than the moving coil type and offer superb reproduction quality.

MAKING YOUR CHOICE

Another factor about headphones which makes them so similar to speakers is that they are highly personal. No one can tell you which are the best. All you can do is check that the frequency response and sensitivity is adequate, and then listen to them. If you like them, get out your cheque book (or is it credit card?) and sign away some more of that hard-earned cash. Still, if you can't spend it, what is it for?

An often overlooked factor in choosing headphones is wearing comfort. It is all very well being over the moon with the bass response, but you will have an unwelcome response from your outer ears half way through a symphony if the headphones do not sit well. Make sure they are comfortable before you walk out of the shop.

Of course, many are adjustable, and this might do the trick. But, if it does not help, then look for other equivalent quality sets which match your ears!

Choose your headphones and your speakers well, and you will be in for hours of enjoyable listening.

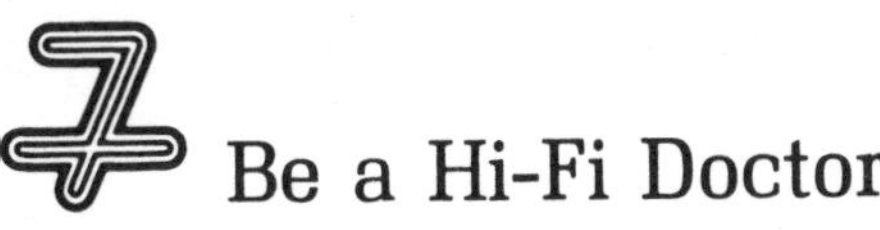

7 Be a Hi-Fi Doctor

One of the frightening things about Hi-Fi equipment is that it appears to be extremely complicated. A single glance inside any unit at the mass of wires, transistors, diodes, and other components sets your head spinning.

It is completely understandable why many Hi-Fi enthusiasts look on quite helplessly when their equipment malfunctions. They just blink at it unbelievingly, switch it on and off a couple of times—perhaps even wriggle the wires around—and then give up in despair. They know that even if they open the machine up, they will not know where to start looking.

So, off they go to the Hi-Fi repair shop.

In actual fact, it could be something quite simple. Something that, if you knew how to identify it, you could fix with little difficulty.

That is what this chapter is all about. It helps you to diagnose the fault, isolate the problem, and provided it is not a major task, how to repair it. Follow a few well-tried steps and you will save yourself pounds in repair bills.

WHERE TO START

The first and most logical place to start, as would any doctor in training, is to make a study of the anatomy and physiology of electrics and modern electronics. A basic understanding of how to keep the power supply to your

Hi-Fi outfit under control, and of circuits and circuit diagrams is vital to our task so that even when confronted with a new piece of equipment, with a circuit diagram to hand, you will be able to find your way around.

POWER TO THE OUTFIT

Attention to the basics—to the simple things in Hi-Fi—is what will minimise the possibilities of faults occuring. You may well have been fitting plugs to electrical units for years, and this portion of the chapter may seem very basic to you. But, it will do no harm to review the business of power supply and see if your technique is skilful enough to ensure safety to you and your Hi-Fi outfit.

Always treat electricity with absolute caution. You cannot see it but it has a nasty bite. Sometimes it is fatal. So, make a standing rule to switch off before you even lift a screwdriver to work on any part of your equipment.

There are three fundamental rules to make sure that the correct amount of power reaches your outfit and to secure all the safety factors necessary to prevent damage.

Rule 1. Be certain that your main plugs are correctly wired. Commit the colour code to memory and then there will be no chance of making a mistake. Remember, it is brown to positive, blue to negative, and green and yellow to earth.

All mains plugs must be wired like this. If you reverse the blue and brown wires, for instance, it could make the chassis of the unit to which it is being supplied live.

Whilst on the subject of mains plugs, always make sure you screw the connections up tightly and do not trap any of the wire's sleeving around the terminal post. Do a tidy job and you can be certain of a high safety factor.

Rule 2. Use the correct fuse. When you buy a plug, a fuse will be fitted. If it is the correct one, always replace it with one of the same type. However, do make sure it is the correct one. You do not need a 13 amp fuse for everything.

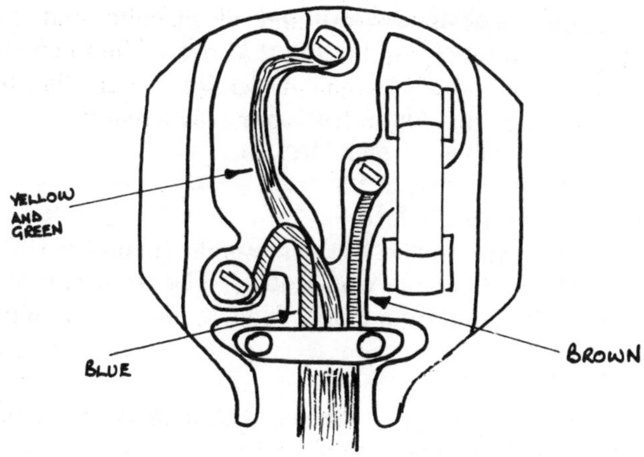

Fig. 37 Three-pin plug showing wiring.

It is easy to work out what you will need. In fact, we have already touched on it in Chapter Two. We talked about voltage and amperage and the formula to find wattage. Well, you can work out the missing factor in any direction. For example, if you know the wattage and the voltage, the amperage is not too difficult to calculate.

You can remember how to work out the calculation if you visualise this: $\frac{W}{VA}$ (W = watts; V = volts; A = amps).
To find the calculation for the missing 'element', you simply cover it up. For instance, if you want to know the voltage, you cover up the 'V' and you are left with $\frac{W}{A}$. Therefore, you divide the wattage by the amperage.

Now, isn't that simple?

But, what good is this calculation? Let us see. We want to know what fuse to fit to a plug. To begin with, we know the mains voltage—it is 240 volts. A check reveals that the item to which we are supplying power consumes 1kW (1,000 watts). Covering up the 'A' in our little formula shows us that we must divide the voltage into the wattage. By our figures that is 240 (volts) into 1,000 (watts). The answer is 4 (amps).

From this it is obvious that to use a fuse rated at 2 amps would be insufficient. On the other hand, a 13 amp fuse is too much. Fuses are supposed to be the weak links in a circuit and 'go first' before any serious damage is done. You can see that a 13 amp fuse in a plug where one of only 4 amps is needed is far from weak. Something quite nasty could happen before the 13 amp fuse 'blew its top'.

In practice, you should fit the nearest fuse above the one calculated. Our example was 4 amps. The nearest one you will find to that will be 5 amps. That is the one you should fit.

Rule 3. Check that the voltage selector on your Hi-Fi unit is switched to the correct one. Many units you see in the shops originate from overseas. They have been produced to function in any one of several different countries. Therefore, they have a selector switch to allow you to adapt them to the 'indigenous' voltage supply.

Your instruction book will tell you where it is. Locate it, and in the British Isles, switch it to 240 volts. We would not give much for your chances of being able to hear any other sounds emitting from your set except a bang if your selector is pointing to anything less.

DOUBLE CHECK

You cannot treat electricity and mains supply carefully enough. It will pay you to double check everything after you have fitted a new plug, or fuse, or anything. A quick run through of all power fittings when cleaning would not go amiss either.

If anything goes wrong and you survive (yes, it is that serious!), the damage to your equipment (and possibly, your home) could be extensive and expensive. In fact, it might even be irreparable. If that happens, all the Hi-Fi 'first aid' you ever learned will be useless.

Remember, the stakes are high. The object of the exercise is to go on enjoying your outfit for many years to come. So, be careful.

CIRCUIT ELEMENTS

Now let us turn our attention to circuits and how to understand them. Any circuit has two basic elements—the input (or energy source), and the output (or component to which that energy is supplied). In between, you will find a variety of other elements. They are designed to take the input signal, control it and shape it up and eventually convert it into whatever is required.

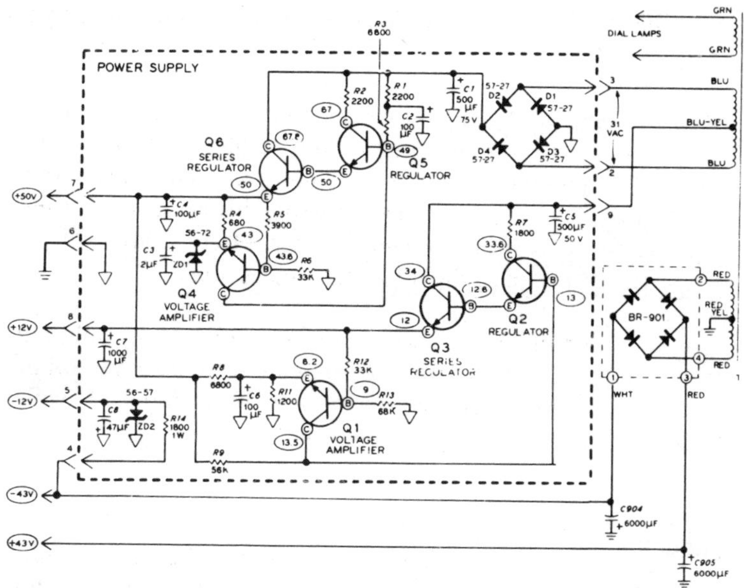

Fig. 38 Circuit diagram of the Heathkit AM-FM solid state stereo receiver, model AR-1500A (power supply circuit only).

All of these components appear on circuit diagrams, of course. But, if you were to detail full descriptions of each element, the diagram would be pretty jumbled. So, for simplicity, and to make the diagrams less taxing on the eyes, symbols have been devised for these elements.

READING CIRCUIT DIAGRAMS

You will already be familiar with the names and purpose of some of these components, but to help you consolidate and advance that information, we will list them together with their symbols.

Component	*Symbol*
Input/output jack sockets	
Resistor	
Variable resistor (potentiometer)	
Pre-set variable resistor (potentiometer)	
Transistor	
Capacitor (or condensor)	
Variable capacitor	
Electrolytic capacitor	
Rectifier	
Diode (or half-wave rectifier)	
Mains transformer	
Fuse	
Switch	
Inductance	

Now let us take them one by one.

Input/output. This is self-explanatory and was listed merely to make you familiar with the symbol. Watch for that. The jack plug symbol is easy enough, but it may be shown with a positive or negative sign alongside, or even the words.

Resistor. In any form, a device for controlling the current and is measured in ohms. The values can be anything from a few ohms to many megaohms (millions of ohms). You will note that some of them are variable, indicating that they

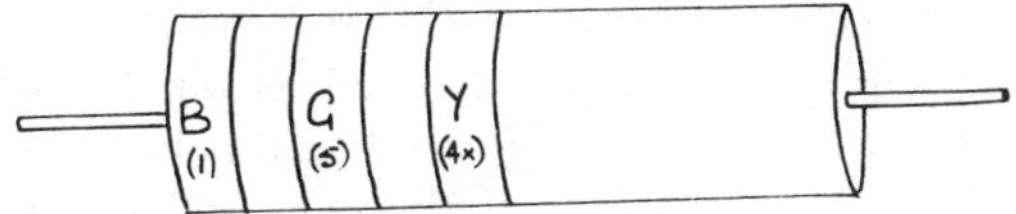

Fig. 39 A resistor with colour coding.

are adjustable throughout all the values within their ranges. This type of resistor is mostly used in the form of volume and variable tone controls, and is also known as a potentiometer.

Colour coding resistors. There are a great many types, sizes and values of resistors. However, they are all numbered and colour-coded for easy identification so that when you see them in 'real life', you can easily spot their values.

0 = Black	5 = Green
1 = Brown	6 = Blue
2 = Red	7 = Mauve
3 = Orange	8 = Grey
4 = Yellow	9 = White

To read the colour code, you first note the colour nearest the end. This tells you the first resistance figure. The next colour tells you the second and the third tells you the decimal multiplier.

Let us look at an example. The resistor we are looking at is coloured brown, green and yellow. By checking our chart we see that brown is 1; green is 5; and yellow puts the decimal point 4 stages over—in other words, we add four noughts. Therefore, our example resistor has a value of 150,000 Ω (150 kΩ).

Transistor. This is a semi-conductor which, among other things, amplifies the signal.

It is because of the introduction of the transistor to domestic electronic equipment that electronics have made the greatest advances in recent years. To appreciate that, let us study its function in an amplifier to see what makes

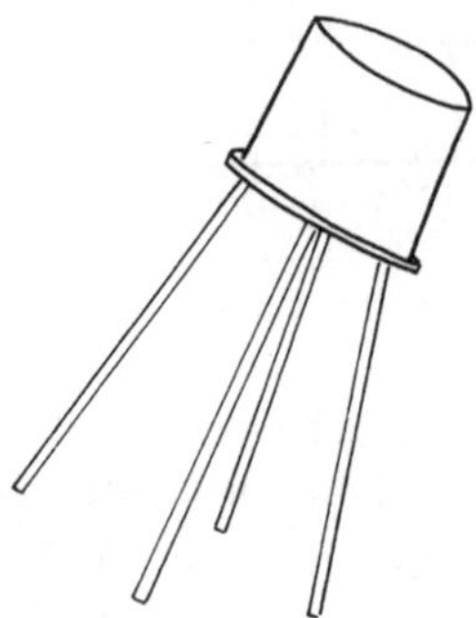

Fig. 40 A transistor.

it tick. (Only a euphemism. If it does tick, switch it off. If it still continues to tick, report it to the bomb squad and stay well clear!)

All amplifier circuits can be divided into three sections or stages:

1. Input stage
2. Equaliser stage
3. Driver and output stage

1. The input stage, or section, is where the minute signals generated from a pick-up, tape head, microphone, etc. are received and boosted. The signal has to reach a level high enough to be handled by the next stage.

2. The equaliser stage is where the signal is 'shuffled' into shape and the bass and treble response is modified to suit the programme material.

Components in this part of the circuit are one of two things—either passive or active. Passive components are capacitors and resistors arranged to modify the frequency response while attenuating the signal as it passes through the circuitry.

Active circuits are more common. They have transistors in the circuit of capacitors and resistors and actively modify the necessary frequencies as required.

The passive system has lost ground to the alternative because further amplification is required to rectify the energy loss incurred. This additional amplification is not

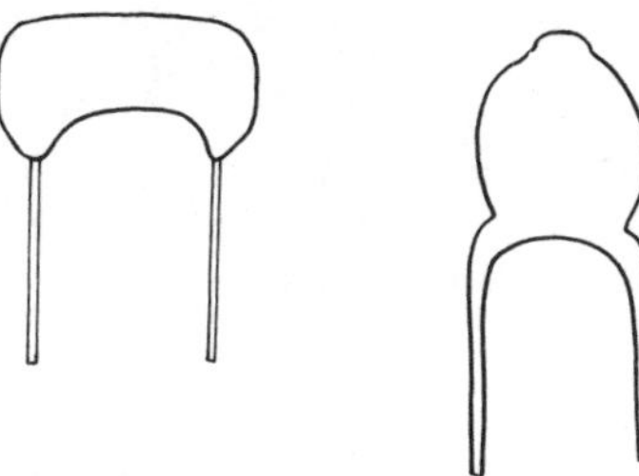

Fig. 41 Two types of capacitors.

needed in the active circuit, and the signal can be passed immediately to the next stage.

3. The driver stage amplifies the signal sufficiently to drive the power output transistors, and from here, to the loudspeakers.

Capacitor. Sometimes known as a condenser, it has the function of storing energy, and is constructed of one or more pairs of conductors separated by an insulating material. Some capacitors can be so constructed that they store a fixed charge. In practice, capacitors are an open circuit to DC voltage, but they impede AC voltages at certain frequencies.

Variable capacitors are able to handle any capacity value within their respective ranges.

Electrolytic capacitor. These are constructed slightly differently. The insulating material between the plates is a thin film of metal oxide. These are formed on the plates by the action of an electrolyte. Its special construction means that it is polarised and must be connected positive to positive, and negative to negative. This type of capacitor provides a high capacity in a limited space and is used for storing electricity and for filtering AC ripple (a small AC component of a direct current).

Rectifier. This converts an alternating current to a direct current. The half-wave rectifier, or diode, does the same job and the current flows in one direction only.

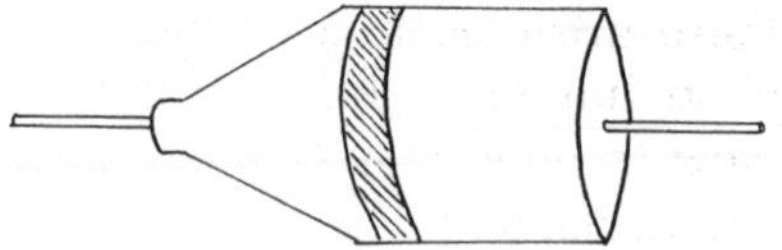

Fig. 42 A diode.

Mains transformer. Used as a means of adjusting the voltage to a determined amount. In the case of transistor amplifiers, the voltage required ranges from about 9 to about 50 (DC). The supply voltage is 240 (AC). In a nutshell, the job of the mains transformer and the rectifier is to make the latter voltage supply become the former.

Fuse and switches. Everybody knows what they are and what they are for. However, you may not have known the diagrammatic symbols. Study them and remember, for you will see them frequently.

Inductance. The property of a circuit which creates a magnetic field and stores it as the current passes through. Inductance is the measurement of that magnetism. The component designed to have inductance is not unnaturally called an inductor. Sometimes to increase the inductance, a coil is made around a metal core (signified on the circuit diagram symbol by a line alongside). Inductors are used to keep the alternating current flowing regularly, particularly during the 'back flow' stages of each cycle.

BUILDING YOUR OWN

Now that you can read a circuit diagram, we do not really recommend that you begin to construct your own Hi-Fi units immediately you have read this book. You will need some experience of Hi-Fi, and a lot more information about circuits than this book is designed to give you before you start.

It is one thing to put a unit together, but quite another to be able to trace the circuit through to find out why it does not work when you have finished.

However, there are companies which manufacture kits and they can be fun to build. You get full instructions showing you how to build step by step, and you will gain considerable knowledge and experience from doing so. Then, of course, there is the satisfaction of having built it yourself.

If that sort of thing 'turns you on', shop around for the kit you want. Hi-Fi magazines usually have advertisements of the different companies wanting to sell their Hi-Fi wares in kit form.

FINDING YOUR WAY AROUND

A little knowledge of what goes on in an electronic circuit does help you to find your way around when something goes wrong. Of course, faults in the circuitry itself, are not something for the beginner to handle. A lot of very expensive and complicated equipment is needed, used by someone with an extensive knowledge of electronics, to effect repairs of today's Hi-Fi equipment.

However, provided you can locate the fault, and it is a simple job, there are some things you can do. It does not require a specialised knowledge to handle a few of the more common simple problems, and there is no reason why you cannot have a go. The important thing is to be sure you know what to do before you begin.

GATHERING TOOLS

Well now, budding Hi-Fi Doctor, you are going to have to collect a few implements—the tools of the trade.

The first thing you should consider is a soldering iron. There are several different types available, so you will have no trouble finding one which suits you. The man who is doing this kind of work every day probably has two or three types—the main differences being in size.

He will use a small 15 watt iron for the general circuitry work. A large 35 watt iron would be needed for such things as heavy duty earth tags, and another (which is probably the best compromise for the amateur) is a medium sized 25 watt iron. This will cope with some of the larger jobs, and with care, will also be alright for some of the more delicate operations.

You will need some solder, of course. The multi-core type is best. This is a lead solder with a 'flux' built in. While you are at it, you might as well buy some desoldering 'braid' too. It is a valuable aid for removing components (you will learn about both of these items later).

Pliers and strippers. Next, you will require a pair of long-nose pliers with insulated handles. A charge of 240 volts (AC) up your arm can be lethal. Even with the mains turned off you cannot be sure of safety. Remember that the capacitor can store electricity and it is quite willing to share it with you through a pair of un-insulated long-nose pliers.

Wire strippers are extremely valuable. They do not have to be expensive. There are some very efficient low-cost ones around. In the same vein, insulation tape is a must for your tool kit.

The multimeter. Finally, stepping up into the realms of the professional electronics engineer, you might consider a multimeter. This instrument is used for measuring AC and DC voltages, currents and resistances. It is a useful thing to have and can be used for a variety of purposes other than Hi-Fi. For instance, it will aid you in finding out if there is a contact between any two or more metal objects in the home, in the car, and so on.

Using a multimeter is not difficult. You will find a range for AC voltage and a range for DC voltage. Whichever one you are working on, always start with the highest point in the range and work down. You can easily render the meter useless if you select 10 volt AC and put the probes into 240 AC.

To measure a fixed resistor, set the meter to the ohms resistance scale and position the probes at each end. Then read off the scale. As you have learned your colour codes, you will know if it is correct.

You can measure an electrolytic capacitor from either

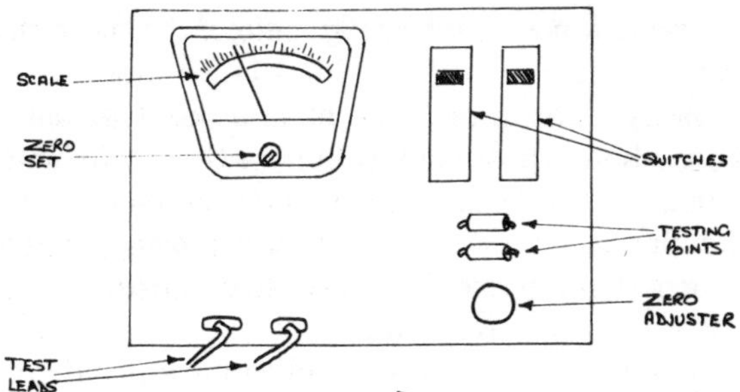

Fig. 43 The multimeter.

end. If it is still charged with electricity the needle will shoot over to maximum, then gradually settle back. If the capacitor is good, it will measure a high resistance. If the resistance measured is low, then throw the capacitor away.

Rectifiers (or diodes) can be measured in this way too. A good rectifier will show a high resistance one way round, and a low one the other way round. If the multimeter reads high or low for both ways, then the rectifier is useless.

A multimeter is not vital for the beginner, but it is a good investment as you become more experienced in sorting out your Hi-Fi problems. When you are ready to purchase one, shop around until you find one which will do the job at the price you can afford. You will be surprised at how inexpensive they really are.

DIAGNOSING FAULTS

O.K. You know something about circuitry components and something of what they do. You have your tool kit standing by and you are ready to be a Hi-Fi Doctor. Then it happens. Your Hi-Fi outfit gets sick. You know it is sick because it is either doing something it should not, or it is not doing something it should.

This is it! Tool kit in hand, you rush to its aid and . . . but, where do you start? Nothing is more irritating than to know something is wrong, but not to know exactly what it is and how to go about putting it right.

Diagnosing a fault is not really difficult. The trick is to localise the fault by identifying what is working properly.

Take this as an example. You have a 'hum' problem (we are talking about Hi-Fi, not B.O.). When you switch on your equipment, a hum emanates from both speakers. Using the process of elimination, you can rule out the loudspeakers themselves. They will either work, be silent, or distort. They will never hum.

That is one link in the chain ruled out.

ISOLATING THE FAULT

The next step is to switch off the power and disconnect the auxiliary units (tape and record decks, tuner, etc.). Then turn on power to the amplifier alone which is only connected to the speakers. If the hum is still there, you know that the fault lies in the amplifier.

By all means, turn off the power again and have a look around inside. It may only be a wire hanging loose. But, if everything appears to be alright, the chances are the fault is in the power supply, or perhaps it is caused by the breakdown of a transistor. Well, if that is the case, until you become more experienced, forget it. It has become a job for the expert.

If, on the other hand, when you check the amplifier by itself, the hum has gone, you know the fault lies in one of the other units. Systematically reconnect them to the amplifier one by one—always switching off the power first. Sooner or later you will plug in the unit which is causing the trouble.

The great advantage with isolating a fault in this way is that you do, at least, know which unit to take to the engineer. Certainly, it may be something you cannot fix yourself, but as you study the subject more, and become experienced, you will be able to do more and more by yourself.

Nevertheless, it is surprising how many silly little things cause what at first appear to be insurmountable problems, but are easily rectified. If you can locate the problem, using the technique outlined above, you can save yourself much frustration, time and money. After all, you can find out if a plug is making proper contact just as easily as a professional electronics engineer at £5 or £6 a throw.

For instance, if your equipment hums, it may not be something expensively wrong with any of the units. It may simply be an earth loop.

'SEPARATES' LINKING PROBLEMS

This problem occurs frequently when you start linking Hi-Fi 'separates' together. However, it is not a difficult situation to overcome. It is the result of two or more units connected together, each having its own earth return. If the earths are connected when linking the units together, then another earth path is created.

The result is an audible and annoying hum.

Leap to your outfit's rescue, Hi-Fi Doctor, and make the amplifier—the heart of the system—the central earth point. Earth it through the mains, and any auxiliary units can then be 'screen' earthed (a screened cable is an earthed cable).

That way, all earths are at one common point (the amplifier) and, the result is, no hum.

CHECKING FAULTS

So, let us now have a look at some of the more common faults and how to cure them. Some of them may be very simple and obvious. But, have patience. You would be surprised how many Hi-Fi outfits have made the long journey back to the sound engineer simply because the input and output plugs were in the wrong way around. Do not be caught out by things such as this. Before you trek to the engineer, check the problem out thoroughly and save yourself a whole lot of trouble.

To aid you in this, use this Fault-Finding Check-List.

Symptom	*Possible Cause*	*Cure*
Equipment active but speakers dead.	1. Speaker wires disconnected.	1. Re-solder or re-fit plugs.

Symptom	*Possible Cause*	*Cure*
	2. Speaker overload fuses blown (if fitted).	2. Replace fuse.
Equipment inactive.	Mains power not reaching equipment.	Check for loose wires or faulty plugs. Replace fuse.
Equipment active but no signal except 'hiss' from speakers.	1. Bad connections from auxiliary units.	1. Check connecting plug.
	2. Faulty cartridge.	2. Check pin contacts on cartridge. (Use small screwdriver on cartridge contacts at low volume. Should emit hum from speakers. If not, have cartridge checked.)
	3. Unit disconnected from amplifier.	Reconnect.
	4. Wrong mode selected.	4. Select correct mode.
Unclear sound from record.	1. Wrong, or damaged stylus.	Engineer will examine stylus under microscope. If necessary, replace.
	2. Arm incorrectly set for weight and bias.	Adjust correctly.

Symptom	*Possible Cause*	*Cure*
Unclear sound from end of record only.	End of side distortion.	Check arm adjustments. If distortion persists, save up for a new arm.
Crackling noises when operating volume or tone controls.	1. Dirty carbon tracks in the potentiometer.	Spray on component cleaner (obtained from dealer).
	2. Badly worn tracks.	Have potentiometer replaced.
Low output and/ or distortion.	1. Pin connections on cartridge corroded making bad contact.	Remove and clean. Keep contacts separate.
	2. High frequency speaker units faulty.	Have replaced.
Rumble from turntable.	1. Dirt or lack of lubrication.	Remove platter, clean and oil in accordance with instruction book.
	2. Wear.	Have overhauled and worn parts replaced.
Wow and flutter in tape recorder.	1. Tape slip due to build up of oxide particles on pinch wheel and/ or capstan.	Clean thoroughly with methylated spirit.
	2. Worn parts.	Have overhauled and replace parts.
Noisy recording even with new tape.	Heads and/or tape path magnetised.	Have all tape path demagnetised.

Symptom	*Possible Cure*	*Cure*
High frequencies attenuated.	1. Heads dirty.	Clean with methylated spirit.
	2. Tape twisted.	Re-insert correctly.
	3. Record and/or playback heads azimuth mis-aligned.	Have re-aligned.
	4. Heads badly worn.	Have replaced.
	5. Bias and/or record equaliser wrongly adjusted.	Have re-adjusted.

This list is not exhaustive. There are many more faults which you will encounter. However, these are the more common ones to watch out for.

SOLDERING

Many of the repairs you have to undertake, or even just putting your units together, will have to be done by soldering. Many tyros think that soldering is easy—until they try it, that is. Then they realise that there is more to it than meets the eye.

After some practice however, soldering does become a relatively simple matter. But since there are no courses available, it is going to have to be a learn-it-yourself process.

There is a right way and a wrong way of going about soldering. Unfortunately, the wrong way turns up more often than is desirable. Many an engineer has wasted hours checking suspected components in a circuit only to find that a bad solder joint is the root cause. However, some of the best engineers fall flat when soldering, so you will be in very good company.

But, do not despair. Digest the information in the next few paragraphs and there is no reason at all why you should not be able to solder as well as any wireman.

IN THE PAST

Back in the good old days, soldering was done with an iron which was usually a huge lump of copper on the end of a thing which looked like a poker. This had to be heated up in the fire or on a gas burner for what seemed ages. A stick of lead was the solder and a tin of brown, greasy looking muck called 'flux' was used to assist the flow. And soldering was hell.

Today, we do not have to go to all this trouble. This is largely due to the advent of the electric soldering iron, and cored solder (lead impregnated with flux). The electric irons provide almost instant heat and, at worst, take a minute or two to reach full, controlled heat.

BEGINNING TO SOLDER

Before attempting to solder, gather your tools around you. You will need a pair of cutters, and your long-nose pliers. Both must be insulated, of course. You will also require a screwdriver, a razor blade or sharp knife, desoldering braid, and a small file.

Cleanliness is the foundation of good soldering. To begin with, switch on the iron and file the 'bit' back until it becomes bright. Then, before it tarnishes again, melt some solder on the bit and shake off the residue. Now your iron is ready for use.

But, that is only half of it. The joints to be soldered must also be clean and free from grease. If they are tarnished, then scrape them with a razor or knife until a bright surface is visible. Each joint to be soldered must then be 'tinned' by melting some solder onto them.

Finally, you place the joints together and apply the iron together with the solder to them. Enough solder should be melted onto the joints (not onto the bit!) to encapsulate them. When you think that sufficient has been applied, remove the iron and allow the joint to cool without moving it. If you manage to dislodge the joint, even slightly, before the solder has set, you may get what is called a 'dry joint'. It will look alright on the outside, but may be an open circuit between the components soldered.

A HEAT SINK

Handle components carefully when soldering them. They are easily damaged, especially by the heat of the iron. You can avoid this by utilising a 'heat sink', or heat absorber. Professionals use a small clamp, but a pair of pliers will do the same job. They must be positioned between the joint being made and the component body. The pliers then, simply attract the heat into themselves and thus, away from the business end of the component.

Being so fragile, components are easily damaged if you have to remove them from a circuit board. To make the job easier, you will literally have to desolder it. A desoldering tool, or desoldering braid are the things to use.

Desoldering braid is a flat coaxial braiding impregnated with flux. You position it on the joint to be desoldered and place the heated iron on it. The solder, as it melts, will flow and be absorbed by the desolder braid, allowing the components to be safely disconnected.

WORKING WITH SOLDER

You may have to solder a variety of things. It is not only confined to internal components, but also applies to one of the most common items which will require the attention of a soldering iron—the plug.

It is unfortunate that there are so many different types of plugs and sockets. This has happened because manufacturers had their own ideas about what was best, or simply used what was available at the time. Then, some were imported from the Continent, and the whole thing got out of hand.

Happily, the situation is now settling down and we are now seeing some standardisation on the horizon.

DIN PLUGS

But, let us have a look at the most common plugs you are likely to find on various Hi-Fi units today and how they are fitted. The first is the DIN speaker plug.

To fit this kind to the cable, begin by sliding the outer casing onto the wire. If you do not make this the first action, you will find yourself in the position of having to disconnect the plug after it has been neatly wired because you left off the outer casing. That can be annoying.

Prepare the cable ends by stripping back some ¼-in. of the outer PVC covering. Then 'tin' the bare ends of the wire and the plug connections and solder in the way you have already learned. Be sure to keep the wiring the same on both channels (or all four if you have a Quadraphonic outfit)—the centre, flat pin is used for the negative and the round pin is the positive.

Check too, that you have left no hair wires sticking out, and make certain that the outer sleeving fits right up to the terminals. If it does not, these two wires could touch and the result of that would be blown amplifier output stages.

If you have to run the speaker cables a considerable distance, you will get better results by using a heavier duty type of two-core cable, rather than the normal bell wire. This is because, when bell wire is run over any distance, attenuation can occur. It acts as a resistor.

Some amplifiers fit 4 mm (banana) plugs for loudspeaker outputs. These are very easy to wire in that they are colour coded. Red is for positive, and black is for negative. This type has wire-retaining screws just like you would find in any ordinary household plug. However, you would be advised to 'tin' the bared ends of the wire. The screws are sometimes sharp and tend to cut through the strands of wire. 'Tinning' prevents this happening and ensures a good contact.

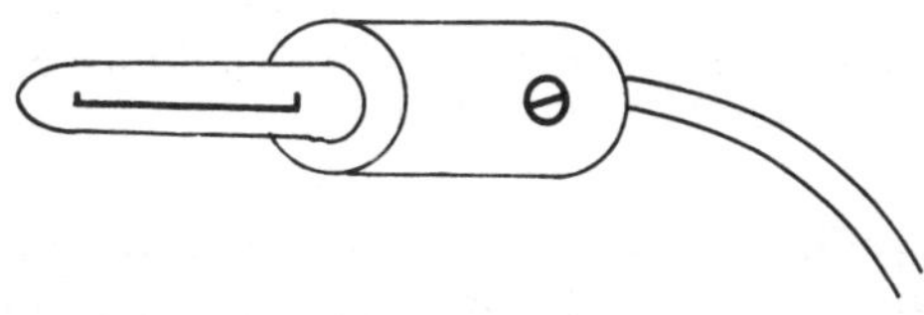

Fig. 44 The banana plug.

SCREW TERMINAL SPEAKER PLUGS

The other type of loudspeaker connector is the screw-terminal plug. This type is probably the best as it is not easily pulled out and disconnected like the DIN or 'banana' plugs.

The best way to connect the cable to screw-terminal plugs is with solder tags, or 'spades'. These are flat metal connectors with holes in them through which the screw secures to the terminal. Some have been designed for a quick release. The hole is a 'U' shape, with the ends of the 'U' reaching the edge, instead of a complete circle. This eliminates the need to completely remove the screw when disconnecting.

You can, of course, affix the wire directly with the screw. This is quite a satisfactory arrangement, if done correctly, when the installation is a permanent one. Fixing it is a simple matter. All you do is to strip back about ½-in. of the PVC covering, twist the bare wire and wrap it around a small screwdriver, remove it, then solder it in that shape.

INPUT PLUGS

So much for speaker connections. Let us now move on to the input side. Here again, the DIN plug is very popular. However, this time you are going to have to be exceptionally careful in your soldering operation. The pins on this type of plug are very close together (there can be as many as five) and a small slip could cause serious damage.

DIN plugs all have pin numbers and your instruction book will tell you what the connections are. However, in the meantime, DIN specifications are as follows. For the turntable input, the positive is the left pin, number three and right pin, number five. Pin two is the common earth for the cartridge. On the recorder input on the amplifier, the DIN socket/plug is used to send the signal out to the tape recorder and pin number one is the left stereo channel and pin number four is the right channel.

The same socket is used to carry the signal from the tape recorder, or tuner, or whatever, back to the amplifier

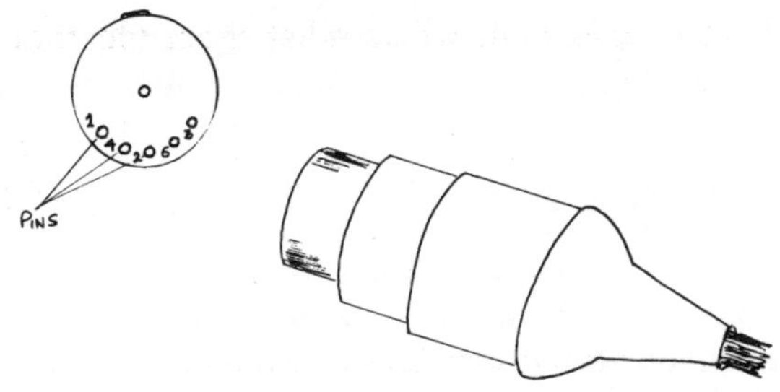

Fig. 45 The DIN plug.

input. In this case, pin number three carries the left stereo channel, and pin five carries the right-hand channel signal. Again, pin number two is the common earth.

On the tape recorder itself, pin one is for the left channel input and pin four carries the signal to the right channel. Pin three is the left channel output, and pin five the right channel output.

THE TUNER OUTPUT AND INPUT

If the tuner output is of the DIN type, then pin three is for the left channel and pin five for the right. The input, on the other hand, is usually a co-axial television-type aerial plug. For cabling, the aerial down lead should be carefully stripped of about 1 in. of its outer sheathing.

Then, insert a small screwdriver into the braiding and, working around the cable, 'comb' it out until you reach the outer sheath. Fold the now straight braiding back over the outer sheath and trim back to about ½-in.

Carefully cut back the inner sheath by ¾-in. and twist the bared inner wire. 'Tin' it and slide the outer threaded locking ring and cable grip onto the aerial wire. The next step is to slide the inner polythene connection onto the inner aerial wire and solder into place. You are going to have to work quickly and accurately because the soft polythene easily melts. If that happens, you will have to start all over again with a new plug.

Re-assemble the rest of the plug, making sure that the outer braiding is trapped by the cable grip. Then, screw the locking ring up tight. This may appear to be a rather crude connection, but as it undergoes little movement, it seems to suffice.

PHONO PLUGS

Another type of connector is the phono plug. To fit it, you should start preparing the cable in exactly the same manner as you would for fitting a co-axial plug. Then, the braiding should be twisted together and cut to about ¼-in. It must then be 'tinned' and the outer casing of the phono plug filed to bare the metal in preparation for soldering. This too is 'tinned'.

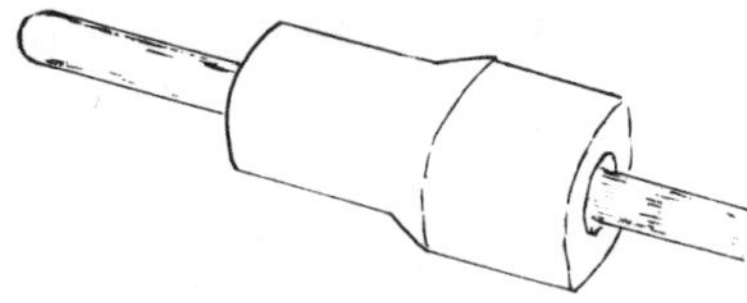

Fig. 46 The phono plug.

Before soldering the joints, make sure you slide the outer cover on first (unless it is the split-apart plastic type). Then, joint the inner wire, followed by the outer wire. Take care that you do not apply toc much solder here, otherwise you will not be able to fit the outer cover over it.

JACK PLUGS

Finally, a connector which is very common in Hi-Fi is the jack plug. This type is usually fitted on head phones. It is relatively easy to cable, the basic process of which has already been covered. However, with stereo, the left channel is connected to the tip, and the right channel to the narrow ring. The longest part of the stem is the common earth. For the mono version, you simply connect the positive to the tip and the negative to the stem.

Soldering and plug fitting are not the most difficult jobs in the world. They do need practice, of course, but provided you follow these step-by-step instructions, you will become proficient in no time. The important thing to remember is to take care. Do not worry about the passing time. It is more important that you get the job done properly. Poor connections due to a rush job, at the very least, will give you erratic signals, at the worst, will ruin your equipment.

Another precaution you should consider is to thoroughly check that you understand what you are doing before attempting to begin. You cannot afford guesswork.

Follow these two simple rules and you will have the makings of a good Hi-Fi Doctor.

8 Caring for Long Life

You have a war on your hands. Unfortunately, it is a war you will have to wage for your entire Hi-Fi life. The enemy is dust. It gets in everywhere and on everything and, if left alone, will cause untold damage.

There are some major areas of concern which need attention. We will explore them in a moment. However, before we reach that point in our strategy, you should institute some elementary protection measures designed to reduce your necessary cleaning to a minimum.

COVER IT UP

To begin with, keep all vulnerable parts covered. For example, the only time your turntable should be left uncovered is when you are playing records. Make sure the heads on the tape deck are not exposed, and never leave your cassette deck lid up. Cassette heads are not the easiest things in the world to get at, so the more you can protect them, the less work you will have to do.

Records and tapes should always be kept in their sleeves and boxes. If you leave them uncovered, you are asking for trouble. This is especially the case when you have come to the end of a playing session. It is very easy to forget the last record or tape at the end of a music session and leave it on the machine.

If you make a habit of this, then persuade your wife to check your Hi-Fi centre before she begins to dust the next morning. Then, if you have forgotten one, she can put it away before any serious damage is done.

Try to break the habit if you can though, and develop one of putting everything away after a session. It is a much tidier and safer habit.

Whilst you are playing records or tapes, do try to keep all covered but the one you have on the machine. Of course, it saves time to have a pile, out of their sleeves or boxes, lined up ready to play. But, it is better to take a little extra time between change-overs manipulating the coverings, than to risk dramatically shortening the life of your valuable music-making collection.

CLEANING UP

However, even if you put into action all of these precautions, you will still have to contend with some dirt. Again, this can be minimised by instituting a regular routine. It is no good merely cleaning your Hi-Fi outfit just before you use it. Dust will be stirred up and things will be worse than if you had left it alone.

You should hold regular cleaning and maintenance sessions. That way you can be sure that your machine is going to function properly, and that dust will be at a minimum.

THE TURNTABLE FIRST

With a soft brush, sweep the dust from all parts of the turntable, including from awkward corners. Periodically, switch off the power and remove the platter. Clean around the inside of the platter and around the outside of the motor pulley with a lintless cloth or a tissue dampened with methylated spirit.

You can take advantage of the exposed turntable machinery to check that all oiling points are clean and that they do not need 'topping up'. If they do, a few drops of fine machine oil will do the trick. Wipe away surplus oil with a clean cloth afterwards.

Finally, when you replace the platter, wipe around the mat with a damp cloth. This is important. If the mat remains covered with dirt and dust, you can bet your life that it will pass the muck on to your records. Conversely, make sure the mat is dry before you place a record on it. Dampness will not help the record surface—nor the stylus when you play the 'damp' side.

LOOKING AFTER RECORDS

Dust is attracted to records by static electricity. There is always some, even in new records, and it builds up the more the record is used. Therefore, when you take a new record out of the sleeve, dust starts aiming itself at it. Then, as you place the record on the dirty turntable mat—well—you can guess the rest.

Of course, you can try wiping the dust off, but what you really need to do is to get rid of the static charge. The problem is not just surface dirt, you see, the muck settles in the groove as well and wiping it away will not easily shift it all. As the stylus tracks the record, it takes the dirt along with it, piling it up on the diamond tip and scratching away at the carefully cut groove.

NEUTRALISING STATIC ELECTRICITY

There is a device available for neutralising the static charge in records. It is called the Zerostat Gun. Shaped like a gun, it is used before you start playing your record. You merely 'aim' the gun at the record, and move it about in a regular pattern over the surface until you have covered the entire disc.

When used properly, the Zerostat Gun will neutralise the static electricity completely, greatly minimising the risk of dust attraction. As static electricity is built up all over again in warm, dry rooms, with lots of soft furnishings (the usual place for Hi-Fi outfits), you can reduce the effect by positioning a bowl of water next to your outfit to raise the humidity.

CLEARING THE DUST

Now unfortunately, when you use the Zerostat Gun, the dust does not fly off. Neutralised static electricity does not mean that the record becomes repellant to dirt. All that happens is, the disc is no longer attractive and loose dust will aim itself at something else.

However, dust which was on the record before you began the process will still be there and has to be removed. This, we must emphasise, also applies to new records.

ROLL OVER

Whether you have used a Zerostat Gun or not, one of the devices for removing surface dirt, practically in its entirety, and a fair amount of groove dirt as well, is the Pixall Drum. This is quite simply a large roller on which there is wound some sticky tape—sticky side out.

At first, it appears to be positively frightening. You will swear that it will make the records sticky too. But, be assured, it does not. As you roll the drum over the surface of the record, all the surface dirt is collected on the sticky tape. The groove dirt is a little more difficult to remove. You need to hold the drum upright and exert a little pressure. Not too much, mind you. You are, after all, rolling a record, not the main street. Even then, the Pixall Drum will not remove all the groove dirt.

However, there is another slight complication. The drum rolling about over the surface of the record tends to build up a static charge. So, in order to get rid of it, you should 'run' the Zerostat Gun over the surface once more before you begin playing the record.

The Pixall Drum does not wear out too quickly. As the tape gets covered in dirt and loses its stickiness, you simply tear off the 'dead' surface, and underneath you will find another brand new layer ready for use.

GETTING IN THE GROOVE

To get at the groove dust you need one of the devices which has a soft felt, brush-like pad and literally sweeps the

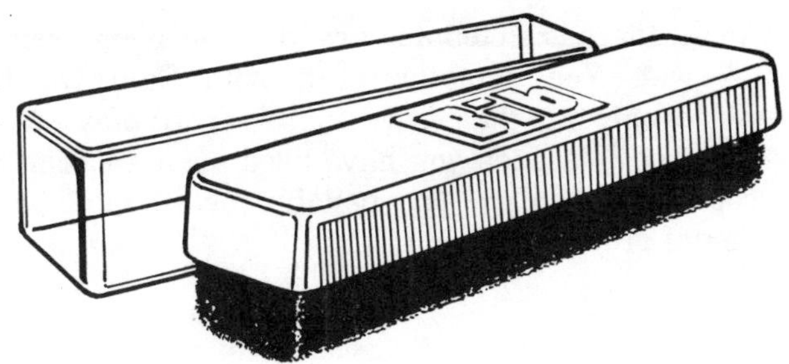

Fig. 47 Bib record 'Dust-Off'.

groove clean. Probably, the best known is the Bib range. But, there are others.

To use, you simply moisten the pad with an anti-static fluid, and sweep away. It is constructed not unlike a clothes brush, so it is quite simple to handle.

An adaptation of it is the multitude of groove cleaning 'arms' which, mounted on their own pivots, track records just like a pick-up. They usually have a small brush on the tip and a felt roller a little further back along the arm which is moistened with an anti-static cleaning fluid and sweeps like the 'brush' mentioned above.

'Moistened' is the operative word. You will have to learn how to get this just right. If it is too dry, then a static charge builds up very quickly. If it is too wet, then groove mud will form. Should your records get in this state, then the only recourse to take in order to restore the record to anything like its former self, is to hand it to your dealer. The chances are he will have a machine for harmlessly scrubbing and cleaning the grooves and neutralising the static charge.

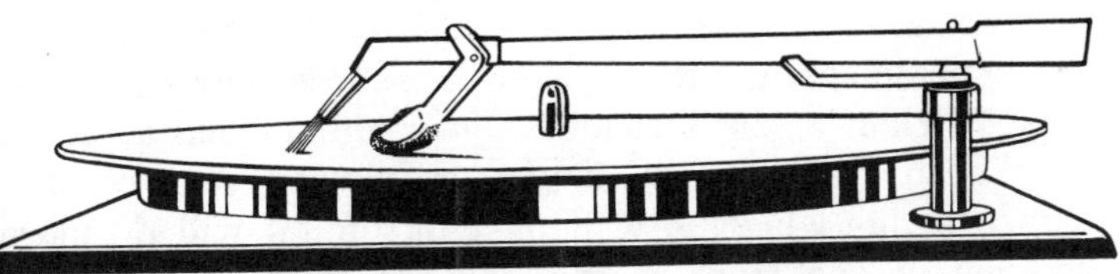

Fig. 48 Bib Groov-Kleen record cleaner.

However, there is a cleaning arm that has been designed to track wet. You actually fill it with fluid which it dispenses evenly into the groove as the record plays. While it does a good job, once you have used such an arm, you must continue to do so. The penalty for stopping is the formation of groove mud.

SUCK IT CLEAN

The ultimate in groove cleaning devices is the Groovac. If you stare at the word a little, you will understand what the machine does. It literally vacuums the groove clean—sucks the dirt right out!

It too works like a pick-up. The pivoted arm is really a bent nozzle with a brush on the tip. It is carefully balanced so that just the correct pressure is exerted on the tip. The motor (a separate unit) is completely noiseless and, when in operation, the nozzle tracks the record, studiously 'eating up' all the dirt as it goes.

Definitely the thing for gadget fanatics, status symbol seekers, and those who merely want their records good and clean. In seriousness, Groovac does an excellent job on all records—old and new. You could do worse than save up for one, even if it will set you back a penny or two.

STYLUS CONTROL

As you learned in Chapter Four, if you look after your records, your stylus will take care of itself. As long as the stylus has a clean groove to track through, it stands every chance of having a very long life—and will, in turn, be kind to your records.

However, do not take too much of a chance on it. Keep an accurate count of the number of records your stylus has played. Then when it reaches the 500th LP mark, get it checked by your dealer. If you try playing your records with a stylus which is well past its life, it will do more damage in one tracking than all the dust and dirt the groove can hold.

STORING RECORDS

Before leaving care of records, we ought to give consideration to storing them. It has already been pointed out that they should be kept in their sleeves at all times, other than when they are being played.

You will no doubt be aware that albums are sold in two sleeves. There is a reason for that. It is not designed to slow you up when you are trying to get at the record. The purpose is to provide double protection against dust. When you put your album away, place it first in the inner sleeve. Then, with the mouth of the inner sleeve uppermost, slide it sideways into the outer sleeve.

The important thing is to avoid having the mouths of both sleeves in the same position. Yes, the mouths together will allow you to slide the record in and out with the minimum of inconvenience, but if the record can get in and out easily, so can dust.

Always store your records fully vertical or, if you must, fully horizontal. Any position in between will lead to warped records. Vertical storing is by far the most desirable method. You would be advised to find a way of keeping them upright at all times, even after some records have been removed.

There are a number of storage racks on the market which do just this. But, if you want to keep them in a special cupboard, build in plenty of thin dividers. It is appreciated that dividers take up space, however, they do ensure that the records will not fall too far from the vertical position when some are removed.

Of course, you may store your albums flat if you want to. But, large piles do tend to put unnecessary strain on the records at the bottom. Long periods in this position could cause damage. The answer is to avoid storing them in this way if you can. Should it be the only method open to you, then make sure that the records are piled on a dead-flat surface (no bumps, or even wrinkles in a cloth) and store them in a series of small piles rather than a sky scraper-like large one.

CARING FOR YOUR TAPE RECORDER

Now let us move on to the tape machine. While it is estimated that the turntable is the most used unit in any Hi-Fi outfit, the tape deck/recorder has the distinction of being the most highly mechanised. It is therefore, more vulnerable to mechanical problems. The cassette recorder, being so miniaturised, is even more vulnerable.

Despite this however, tape and cassette decks and tapes are not so delicate as your turntable equipment. Most of the faults occurring in these two types of tape machines are as a result of mechanical failure rather than anything else. Even then, a good machine—open reel or cassette—will stand up to a lot of wear.

ESTABLISH A ROUTINE

To ensure that your tape machine stays in running order, plan on sending it in to your dealer at least annually for

Fig. 49 Tape cleaning outfit.

checks and, if necessary, an overhaul. In between these yearly trips to the dealers, it is up to you to maintain your machine, to keep it clean, and to ensure its smooth and correct running.

The heads and tape path need priority attention. First, you must sweep the areas lightly with a soft brush to remove deposits of oxide left by the tape. Any stubborn deposits should be removed with a non-abrasive nylon tool. Do not use anything metal. Apart from scratching the surface, the metal could upset the magnetic arrangement in the heads.

A wipe with cotton wool buds moistened with methylated spirit is the next move. Wipe along the tape path to make sure that grooves formed by the passage of the tape are thoroughly cleaned out. In the case of a cassette recorder, the final step is to polish the heads with a cleansing tape. They work just like an ordinary cassette, but do keep an eye on it. When well used it tends to clog with dirt and oxide particles and will then do a good job of scouring even more grooves in the heads.

Particles of dust on the heads and capstans impair reproduction and should be removed frequently. In fact, a light brushing before every session would not go amiss.

Fig. 50 Cassette head cleaner.

Fig. 51 Cartridge head cleaner.

You can have your more serious clean-up sessions at longer, but regular intervals.

DEMAGNETISING

You may find that from time to time the metallic parts of the tape path become magnetised. A recorded tape passing through a recorder in such a condition will quite simply be ruined. The magnetised oxide particles on the tape will be slightly re-adjusted as the tape proceeds on its way and the sound will never be the same again.

However, your recorder will not be beyond further use, a defluxer, available from your local dealer will put the matter right. It will demagnetise the machine, restoring it to normal.

CLEANING TAPES

Tapes need attention too. They tend to store dirt along with the programme content and need to be cleaned on occasions. Fortunately, the protection offered to cassetted tapes is sufficient to ensure their cleanliness. But, open-reel tapes should be cared for regularly.

Cleaning tapes is a simple matter. Merely hold a piece of tissue, dampened with methylated spirit, around the tape whilst it is being rewound. Then repeat the process with a dry tissue. It is important to ensure the tape is not stored wet.

Take care not to touch the surfaces of the tape with your fingers at any time. If you do, moisture will be transferred to the tape causing damage which is difficult to put right.

TAPE STORAGE

Of course, you can alleviate the dust collecting problem somewhat by taking precautions and protecting your tapes. Always store them away in a box, in a dry, warm place. Avoid, if possible, transferring your cassettes from car to Hi-Fi unit. A car is a good place to pick up any dirt going

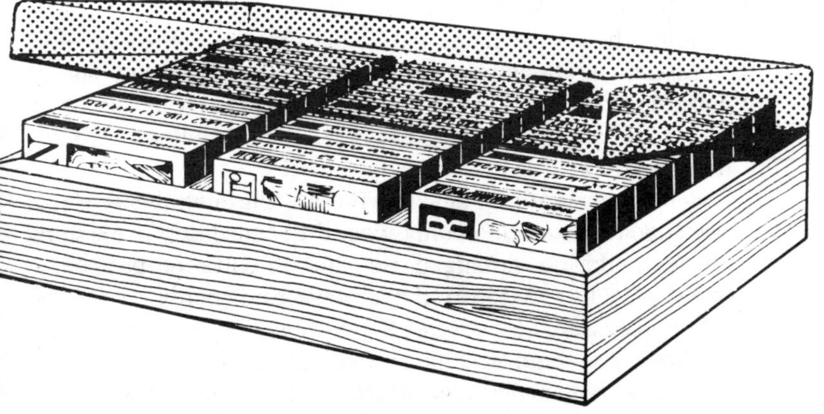

Fig. 52 Bib cassette storage cabinet.

(it is circulated throughout the entire car by the air conditioning system) and your tapes will be among the first things to collect it.

The answer to this problem is simple. Have tapes for use in the car, and tapes for use in the home. Try not to mix them.

PREPARING TAPES FOR USE

Whilst not strictly dust, the effect any loose oxide particles from a brand new tape have on the heads, and in the tape path is the same. After a while, the reproduction quality is impaired. Loose particles can be removed by putting the tape through a fast rewind before recording on it. The effect is such that the loose oxide particles are removed, the oxide surface is polished, and the tape is slightly stretched.

Provided this is done before any sound signals are transferred to the tape, you can be assured of a consistently good sound, and a much better signal to noise ratio.

This procedure is not generally necessary for cassettes, but as a precaution, it will do no harm to go through the motions. However, in either case, do remember to clean the heads with a soft brush before continuing with the recording session.

DISMANTLING CASSETTES

Should it be necessary to get at the tape in a cassette—for cleaning, or because the cassette has jammed—it is possible to take certain models apart. Some cassettes have four small screws holding the two halves together. Taking them apart is a relatively simple matter.

Before dismantling, slowly wind all the tape evenly on to one spool. It must be evenly wound because any projecting edges could be damaged as the cassette is put back together. Then, remove the screws and gently prise the two halves apart.

There is a tape rescue kit on the market which is, in reality, a temporary cassette for holding the tape whilst you are paying attention to its proper home. If you want one standing by, fine. But they are not really necessary except in a few cases. To be on the safe side, ask your dealer before dismantling your offending cassette.

WHAT TO LOOK FOR

As you prise the two halves apart, check the liners. These are teflon discs or 'panels' over the spools. They should not be bent and should allow the spools to revolve freely. Avoid touching the liners with your fingers. Handle them with a tissue—and be careful. They are fragile, and easily bent.

Clean the cassette with a dry paint brush, paying special attention to the inevitable collection of dirt and oxide particles in the tape path. Then, check along the seam where the two halves of the cassette join and remove any burrs that you find. If left, they could scratch the tape.

Finally, put the cassette back together, replacing the liners first (again handling them with a tissue) and making sure they are not bent. Screw the two halves together lightly and check that the spools turn freely. A pencil put through the spool hole and rotated slowly will suffice as an initial test. If you feel resistance and the tape appears to stick, you have re-assembled it wrongly and you should take it apart and begin again.

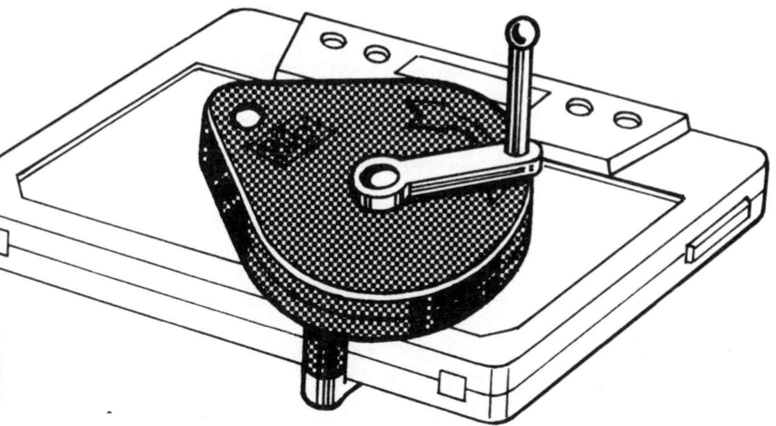

Fig. 53 Bib cassette fast hand winder.

When you are sure that all appears to be well, try the cassette in the machine. If it runs smoothly there, you are home and dry. Tighten the screw properly and then the cassette is ready for regular use.

There is no doubt about it, care for your equipment and accessories and you can be sure of a long life of enjoyable music. Your Hi-Fi outfit is the same as any other collection of delicate instruments, it needs attention and careful handling. So, if you want your outfit to run beautifully for 'forever and a day', tender loving care is your secret weapon in this war you have on your hands. Arm yourself with it now.

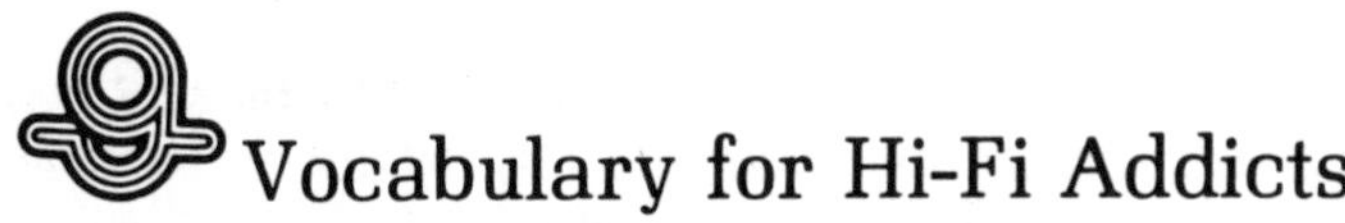

9 Vocabulary for Hi-Fi Addicts

Well, you are now armed with sufficient background information to find your way around the maze of Hi-Fi. This book does not pretend to even begin to tell you which outfits to choose. That is up to you. But, at least you will know where to start looking, and when you find it, you will, at least, be able to understand something about it. No longer will your Hi-Fi choice be surrounded with the mystery that many people consider it to be.

If you have understood everything in this book, you will have no trouble when you start searching for your outfit. If you already have one, then it will have taken on a new light. Already you will be telling the rest of the family to handle your outfit with greater care than they have until now.

However, whether you have an outfit or not, at some time or another you are going to have to enter the shrine of all Hi-Fi fans—the dealer's. It is then that you could experience trouble. As soon as he starts to speak and throw all that jargon at you at high speed, everything you have learned seems to fly out of the door.

BRAIN BLOCK

You know how it goes. There stands your friendly dealer as you enter the shop. He spots you, a smile appears, and he launches himself at you. The moment of truth has arrived.

After the first couple of sentences you slowly sink into a morass of technical terms until you disappear, smothered by a mountain of meaningless jargon. His chatter might as well be a message from Mars.

A FOREIGN COUNTRY

There is no doubt about it, no matter how much of this book you have absorbed, when you venture into Hi-Fi country, you could be forgiven for thinking that you had entered a foreign land. Of course, there are some words you understand, but the meanings of the more important ones seem to pass right over your head.

The same thing happens when you try to read helpful-looking articles in any one of the great many Hi-Fi magazines. Pretty soon you are forced to a grinding halt by what appears to be a lot of mumbo-jumbo, and then, disillusioned, you scan through the rest of the magazine, looking at the pictures.

It is not that the dealer or the magazine editor is trying to out-fox you. It is merely that both of them have 'lived' electronics for so long that they just do not know how to describe Hi-Fi equipment any differently.

What can you do?

LEARN THE LANGUAGE

One answer is to attack the problem as you would when making a visit to a foreign land. It is surprising how you get around 'over there' with a vocabulary of just a few pertinent words.

You can do the same in Hi-Fi country. Memorise a few of the most commonly used words, and a few of the not so common ones, and you will be able to hold your own with any dealer or magazine editor.

However, knowing the words is not quite enough. You have to know a little bit about what they mean so you do not make a fool of yourself and use them in the wrong context, or worse still, in the wrong place in a sentence.

The *faux pas* would be akin to ordering the wine waiter instead of the fish and chips on a French menu.

A CRIB GUIDE

So, what follows is a sort of 'crib guide'. You will find an alphabetical list of Hi-Fi-related technical terms with a short explanation of their meanings. To see any one of the terms in context, you merely cross-reference it with the same word in the main body of the book.

Commit what you can of this list to memory to form the basis of your new Hi-Fi language, of course. But, you will also find it of value to use as a reference for checking on words you encounter in Hi-Fi magazines, equipment reports, etc.

Whichever use you make of this vocabulary guide, it will 'short circuit' things for you as you set foot in Hi-Fi country.

Bon voyage.

ACOUSTICS	The science of sound in all its aspects.
AM	Amplitude Modulated. A sound wave which is added to a radio wave with a constant frequency and a variable amplitude.
AMBIENCE ENHANCING	Circuitry designed to recover ambience from stereo signals.
AMPERE	A unit of an electrical current.
AMPLITUDE	A term expressing the strength of sound.
ANRS	Automatic noise reduction system.
ANTENNA	An American word for an aerial.
ANTI-SKATING DEVICE	A device on a pick-up arm to counteract bias.
ATTENUATION	Another word for the reduction of the strength of a signal by electronic means.
AXIS	An imaginary line which passes directly through the centre plane of an object at an angle of 90° to the front or rear surface.

AZIMUTH The vertical gap in the record/playback heads of tape machines to facilitate recorded tape interchangeability.

BAFFLE Of loudspeakers, a partition designed to increase the effective distance a signal can be transmitted between two points.

BASS REFLEX A term used to describe a type of speaker enclosure which has a port in the baffle in order to increase its effective bass response.

BASS RESPONSE Response to low frequencies.

BEL A unit which describes the ratio of two power values, or signal intensities.

BIAS 1. The natural inward 'pull' of an arc-moving pick-up arm.
2. A VHF signal injected into the recording head of a tape recorder together with the programme signal to improve the high frequency signal to noise ratio.

CAPACITANCE Refers to the property of conductors and insulators which have the ability to store electric charges.

CAPTURE RATIO The ratio of two signals when the stronger has 'captured' or reduced the weaker.

CARDOID Heart-shaped. Of microphones, the shape of the area of sensitivity.

CARRIER A wave which is used in conjunction with a radio wave to increase its transmitting distance.

CARTRIDGE 1. The 'brain' at the end of a pick-up arm which converts the signal from the stylus tracking the record groove, into electrical energy for onward transmission to the pre-amp.
2. A continuous, eight track, stereo, miniaturised tape system.

CD–4	Compatible Discrete Four Channel (Quad).
CHROMIUM DIOXIDE	(CrO_2) The name given to a tape with a high density of magnetised chromium dioxide particles affording very low noise properties.
COIL	Usually, a coil of very fine wire which, when a current passes through, forms a magnetic field.
COLORATION	A word used to describe the inherent electronic influences in a loudspeaker system.
COMPRESSION	The increased air pressure at the peak of a sound wave.
CONE	Specially treated paper or plastic shaped into a cone, electrically induced to vibrate in response to sound signals to move the air in front and behind which carries audible sounds to the ear.
CROSSOVER NETWORK	Circuitry which controls the frequency ranges transmitted to appropriate speaker cones in a loudspeaker unit.
CROSS TALK	The impingement of unwanted audio signals from one stereo channel onto another.
CYCLE	Refers to one complete sound wave. Generally linked with a time scale and stated as cycles per second (c/s).
DAMPING	Energy lost by a vibrating component in overcoming friction caused by air, metal, etc.
DECIBEL	(dB) One tenth of a bel. In general use because the bel is too large a unit with which to work easily. 1dB is the smallest intensity change detectable by the human ear.
DEFLUX	Demagnetise.
DIN	A European standard of measurement.

DISCRETE SYSTEM	Four signals, one on each of four different channels (Quad).
DOLBY	The name given to tape recorder circuitry for the reduction of background noise in recorded tapes.
DUAL CONCENTRIC	A loudspeaker system where the high frequency cone is actually fitted within the low frequency cone.
DYNAMIC MICROPHONE	An alternative name for the Moving Coil Microphone.
EFFICIENCY	Also known as Sensitivity, it relates to the way the speaker cone, when vibrating as a result of the signal input, moves the air to produce an audible sound.
ELLIPTICAL STYLUS	A stylus manufactured with an oval shaped tip in order to conform as closely as possible to the original cutting stylus.
emf	Electro-motive force. The driving force (e.g. a battery) behind a current, measured in volts.
ENCLOSURE	The combination of speaker cone mounting and cabinet which has a marked effect on the resultant sound.
EQUALISATION	1. Refers to the standard established by the RIAA for the control of low and high frequencies in record production and play back. 2. The circuitry in an amplifier which modifies the frequency response.
FLUTTER	Rapid fluctuations of a tape or record speed.
FM	Frequency Modulated. A sound wave which is added to a radio wave with a constant amplitude and a variable frequency.

FREQUENCY	Relates to the number of sound wave peaks which are formed at a given point over a period of time. The greater number formed, the higher the frequency.
FREQUENCY RESPONSE CHARACTERISTIC	A plotted scale which shows how a Hi-Fi unit responds to the range of frequencies it is capable of handling.
GAIN	An amount of increase in signal level, measured as a ratio of input to output and expressed in decibels.
GROOVE MUD	Literal mud which is produced by dust and liquid (e.g. cleaning fluid) in record grooves.
GROUND WAVES	Radio waves transmitted over the earth.
HAFLER EFFECT	Ambience effect derived from linking one or two speakers to standard stereo speakers and positioning them behind.
HARMONIC DISTORTION	The distortion of overtones by the system, measured as a percentage.
HEADS	Name of the electromagnets which transfer the signal to, receive the signal from, and erase the signal from tapes in tape, cassette, and cartridge recorders.
HEMISPHERICAL STYLUS	A stylus manufactured with a circular-shaped tip.
HERTZ	The European equivalent of cycles per second. Now in common use throughout the world.
HIGH PASS FILTER	A network in an amplifier which eliminates or filters unwanted low frequency signals in the programme content.
IF	Intermediate Frequency. Produced in FM receivers by a combination of the input signal and a locally, oscillator-produced signal as a device to increase the signal to noise ratio.
IHF	Institute of High Fidelity.

IMPEDANCE	An amalgamation of resistance and natural opposition to the current (reactance), the units of which are measured in ohms.
INDUCTANCE	The resistance and changing rate of a circuit to the flow of the current.
INERTIA	The continued movement of an object after the driving force has ceased.
INFINITE BAFFLE	A completely sealed enclosure in a loudspeaker which absorbs the sound waves emanating from the rear of the cones.
ips	Inches per second. In the context of Hi-Fi, applies to tape speeds.
JOINTING BLOCK	A device for cutting and holding tapes when editing.
KILO	A prefix to denote one or more thousand. E.g. 1 kilohertz (1 kHz) = one thousand hertz.
LACQUER	The record made from the master tape, so-called because it is an aluminium disc covered with a soft vinyl, or lacquer.
LOAD	The part of the circuit to which the power or signal is supplied.
LOGIC CIRCUIT	Circuitry designed to decode to which channel a signal belongs.
LOGIC CONTROL	A control to improve Quad signals and reduce cross talk.
LOW PASS FILTER	Amplifier network which filters unwanted high frequency signals in programme content.
MASS	Weight distribution in an object.
MASTER DISC	Record produced from master tape—the lacquer.
MASTER SHELL	A 'negative' record produced by 'taking an impression' from the master disc.
MATRIX	Circuitry which decodes stereo and Quad signals into four channels.

MEGA	A prefix to denote one or more million. E.g. 1 megahertz (1 Mhz) = one million hertz.
MICRON	One millionth of a metre.
MILLI	A prefix to denote a thousandth part of a whole. E.g. 1 millivolt (1 mV) = one thousandth of a volt.
MIXER	A unit which allows the production of a single recording from two or more sound sources, either at once, or at different times.
MODULATION	The merging and conforming of one sound wave to another.
MOTHER	A one-sided record produced by 'taking an impression' from the master shell.
MOVING COIL	Of speakers, headphones, and microphones, and relates to a coil which vibrates within a magnetic field in response to the sound signal.
NEGATIVE	Another name for the master shell.
OHM	The unit of measurement of the quantity of resistance, impedance, etc.
OSCILLATOR	A device for producing or converting power or a signal.
OVERTONES	Tones (frequencies) which surround a specific tone (frequency) emitted by a musical instrument or electronic unit, producing a characteristic sound.
PHASE RESPONSE	The speed in which a speaker cone can respond to an input signal and emit an output signal. A characteristic of speaker cones is that they can 'store' some energy signals and not convert them to sound right away.
PHON	The unit of the loudness of a sound vibration or signal.

PITCH	A more common method of describing the frequency of cycles. The greater the number of cycles per second, the higher the pitch.
PLATTER	The rotating platform of a turntable upon which the record rests when being played.
POLAR RESPONSE	The area of sensitivity from which a microphone will receive signals when held in one position, or to which a speaker will transmit signals.
POST-ECHO	A subdued, but audible signal, directly related to the programme content, heard just after the desired signal.
POWER HANDLING	Of speakers, the figure which shows the maximum input the speakers are capable of dealing with.
PRE-AMP	Short for Pre-Amplifier. It can either be separate or a part of the main (power) amplifier. The part of the amplifier where the controls are housed.
PRE-ECHO	A subdued, but audible signal, directly related to the programme content, heard just before the desired signal.
PRINT-THROUGH	When thin tapes are stored unused for long periods, the signals can be transferred from layer to layer and produce post- or pre-echoes.
QS	Quadraphonic Stereo.
QUADRAPHONY	The term given to the production of all-round sound and requires special Hi-Fi units including four matched speakers.
RAREFACTION	The decreased air pressure at the lowest point of a sound wave.
REACTANCE	The result of opposition to a current in a coil.

RESISTANCE — The tendency of the material in a circuit to resist the progress of the current and convert it into heat.

RIAA — Record Industry Association of America.

rms — Root mean square—the square root of the average of the squares of all points in a complete sound wave cycle.

rpm — Revolutions per minute. In the context of Hi-Fi, it applies to turntable speeds.

RUMBLE — Unwanted turntable and/or motor noises amplified with the programme content.

SEL-SYNC — A specialised tape recorder that can play back pre-recorded tracks whilst simultaneously recording on one or more of the remaining, clear tracks.

SENSITIVITY FIELD — The field over which a microphone is sensitive—for example, heart-shaped (cardoid), figure-of-eight, etc.

SHIM — A non-magnetic spacer positioned between two electromagnets.

SIGNAL TO NOISE RATIO — The ratio of the inherent background electronic and/or tape noise to the programme content signal.

SINE WAVE — The form of a complete cycle of a sound wave.

SOUND COLOUR — A term applied to the characteristic sound of a musical instrument or electronic unit, including tones and overtones.

SOUND ON SOUND — A facility in a tape recorder to allow the progressive build up of a complete programme on a single track, one 'take' at a time, by transferring signals from track to track whilst adding subsequent 'takes'.

SQ — Stereo Quadraphonic.

SQUAWKER — The popular name given to a mid frequency-range speaker cone.

STAMPER	The final 'negative' stage of 'master' record production. Used to literally stamp its form in soft vinyl to produce the records sold to the public.
STEREO	A system of transmitting two different sound signals on two different channels to give the impression of 'three-dimensional' sound.
STYLUS	The modern form of a gramaphone needle. Manufactured with a sapphire or, preferably, a diamond tip for Hi-Fi.
SUPERHETERODYNE	The name given to a receiver which combines the incoming frequency with a locally produced one, amplifies it and decodes it for a greater signal to noise ratio.
SYNCHRONOUS MOTOR	A motor which makes use of the frequencies of an alternating current instead of the current itself, resulting in a better speed constancy.
TIMBRE	Another name for a characteristic sound emitted by a musical instrument or electronic unit, including tones and overtones.
TRACKING FORCE	The downward pressure of a stylus tracking the groove of a record.
TRANSCRIPTION DECK	A particularly heavy platter used on some turntables to help to even out any speed fluctuations.
TRANSDUCER	A device which converts mechanical or magnetic energy into electrical energy and vice-versa.
TRANSIENT RESPONSE	The response of a unit to sudden sharp 'explosions' of sound—i.e. sudden inputs of voltage.
TWEETER	The popular name given to a high frequency-range speaker cone.
UD-4	Universal Stereo Four Channel (Quad).
UHF	Ultra high frequency.

VARI-MATRIX	A control which increases channel signal separation.
VOLT	The unit of measurement of the emf.
WATT	The unit used to express electrical power.
WAVELENGTH	Distance between one sound wave peak and another.
WEIGHTED	A standard of measurement which brings the specification closer to the reaction of the human ear.
WOOFER	The popular name given to a low frequency-range speaker cone.
WOW	Slow fluctuations of a tape or record speed.

INDEX

* Indicates that this term also appears in the Vocabulary on pp. 173–83

Acoustic isolation 60
*Acoustics 12–21, figs. 1–7
Aerial *see* Antenna
*Ambience enhancing 85
American Institute of High Fidelity 52
*Ampere 33, 133, fig. 10
Amplifier 27–42, 137–9, figs. 9–12
—— in brief 137–9
—— choosing 23, 28, 31, 41–2
—— controls 28–30
—— matching 31–2, 35, 38–9, 106, 125–7
—— specifications 30–42
—— —— comparing 41–2
—— —— example 31–2
—— tape deck 92
*Amplitude (Acoustics) 20, fig. 7
*Amplitude modulation (AM) 44–5, 47, fig. 13
Antenna 49–51
—— AM 49
—— FM 50–1
—— indoor 50–1
—— input impedance 48, 51
*Anti-skating device 78
Arm 75–80, figs. 21–3
—— components 77–8, figs. 22–3
—— specifications 78–80
—— —— example 78–9
*Attenuation 29
*Axis 125, 127
*Azimuth 148
*Baffle 123
—— infinite baffle cabinet 122
Balance—amateur recording 99–100
—— control 29–30
—— pick-up arm 77, figs. 22–3, 26
—— record production 62–4
—— tape transporter 96
Banana plug fig. 44
—— fitting 151
*Bass reflex cabinet 122
*Bass response 119
*Bel 20–1
Belt drive 69
*Bias—pick-up arm 76–7, fig. 21
—— —— compensator 78
—— tape recorder 95, 148
Bib products
—— cartridge head cleaner fig. 51
—— cassette fast hand winder fig. 53
—— cassette head cleaner fig. 50
—— cassette storage cabinet fig. 52
—— Groov-Kleen record cleaner fig. 48
—— record 'Dust-Off' fig. 47

—— stylus balance fig. 26
—— tape head cleaner fig.49
——tape splicer fig. 35
Cabinets 121–2
—— bass reflex 122
—— home-made 121
—— infinite baffle 122
*Capacitance 105
Capacitor 139, fig. 41
—— microphone 105–6
—— symbols 136
Capstan (Tape transporter) 96, fig. 29
*Capture ratio 48, 54
*Cardoid 106
Care of equipment 157–70
—— cassettes 167–70, figs. 52–3
—— stylus 83, 162
—— tape recorder 164–7, figs. 49–51
—— turntable 157–9
*Carrier 45
*Cartridge (Pick-up) 80–4, figs. 23–6
—— ceramic 80
—— description 81–3
—— importance 81
—— magnetic 81
—— matching 84
—— moving coil 80–1
—— performance 84
—— specifications 81–4
—— stylus 81–3
*Cartridge (Tape) 24. fig. 8
Cartridge deck 90–3
—— & hi-fi 102
—— tape 109
—— vs. open reel 101–2
Cassettes 24, 108, fig. 8
—— care of 167–70, figs. 52–3
—— cleaning 169
—— dismantling 168–70
—— fast hand winder fig. 53
—— storage 167–8, fig. 52
—— tape 108
—— —— speed 101
—— —— tracks 98, fig. 31
CBS Laboratories quad 85
*CD4 86, 88
Channel separation 48, 53, 82, 84
Check-List (Faults) 145–8
Choosing an outfit 22–6
—— amplifier 23, 28, 31, 41–2
—— headphones 128–9
—— kits 140–1
—— pick-up 75
—— quad 87–8
—— speakers 25, 116–7
—— tape recorder 24, 91–2, 96–7, 113
—— tuner 23, 43–4, 47, 55
—— turntable 23–4, 58, 66–7, 73–5
*Chromium dioxide tape 108–9
Circuit—components 136–40, figs. 38–42
—— diagram 132, 135–6, fig. 38
—— simple circuit fig. 10
Cleaning 157–70
—— cassettes 169
—— records 159-62, figs. 47–8
—— tape recorder 165–6, figs. 49–51
—— tapes 167
—— turntable 158–9
*Coil 93
*Coloration 117
Colour coding—mains wiring 132, fig. 37
—— resistor 137, fig. 39
Compatibility 31–2, 35, 38–9, 51, 78, 84, 90–1, 95, 116, 121, 125–7
Compatible Discrete Four Channel (CD4) 86, 88
Components—circuit 136–40, figs. 38–42
—— —— symbols 136
—— pick-up arm 77–8, figs. 22–3
*Compression 15, fig. 3
Condenser 139, fig. 41
—— symbols 136
*Cones 118–21
—— dual concentric 120
—— positioning 120

Controls, amplifier 28–30
Counterweight 77, figs. 22–3
*Cross talk 31, 38
*Crossover network 119-20
*Cycles 16-8
*Damping factor 34-5
*Decibel 21
*Defluxing 166
Desoldering 150
*DIN 125
—— plug fig. 45
—— —— fitting 150-4
—— watts 125-6
Diode 139, fig. 42
—— measuring 143
—— symbol 136
Direct drive 70
Dirt removal *see* Cleaning
*Discrete systems 86-7
Distortion—amplifier 31, 35-6
—— compensation (Arm) 76
—— tuner 48, 53
*Dolby system 103-4, fig. 33
Double tracking 98-9
Drive wheel system 69
Driving power range (Speakers) 126
*Dual concentric speaker 120
Dust removal *see* Cleaning
Dynamic headphones 128
*Dynamic microphone 104-6, fig. 34
Ear 12, 17, 20-1, figs. 1, 4
Earths 145
Echo 100
Editing 109-10, fig. 35
Effective length (Pick-Up Arm) 79
Effective mass (Pick-Up Arm) 78-9
*Efficiency (Speakers) 127-8
Electricity and safety 132-4, fig. 37
Electrolytic capacitor 139
—— measuring 142-3
—— symbol 136
Electrostatic headphones 126
Electrostatic speakers 118-9
—— positioning 124
*Elliptical stylus 82-3, fig. 25
*Enclosure 121
End of side distortion 76
*Equalisation standards 39, 64, fig. 19
Erase head 95-6, fig. 29
Erase ratio 89
Faults 141-8
—— diagnosis 143-8
—— —— check list 145-8
—— isolating 144-5
Filters 28-9
*Flutter 68, 72, 89-90, 96
*Frequency 14-8, figs. 4-5
—— compensation 39, 63-4
—— correction 62-3
—— measurement 16
—— modulation (FM) 44-5, 47, fig. 14
—— radio 45-9
—— range (Acoustics) 16-7
—— —— (Radio) 48-9
—— response 31, 36-7, 39, 48, 53, 82-3, 89-90, 94, 101, 125, 127
—— —— characteristic 36-7, fig. 12
Friction (Pick-Up arm) 78-9
Fuse—calculating required rating 132-4
—— symbol 136
Generating voltage 40-1
Groovac 162
*Groove mud 161
*Ground waves 46
*Hafler effect 85, fig. 27
Half-wave rectifier 139
—— measuring 143
—— symbol 136
*Harmonic distortion—amplifier 31, 35-6
—— tuner 48, 53
Head (Pick-Up arm) 77-8
—— shell 78, fig. 23
Headphones 128-9
*Heads (Tape recorder)
—— care of 165-7, figs. 49-51

— — cassette 102, fig. 32
— — open reel 89, 93-6, fig. 29
— — — — erase head 95
— — — — playback head 94-5
— — — — recording head 94
Hearing 12, 17, 20–1, figs. 1, 4-5
*Hemispherical stylus 82
*Hertz 17, 32-3
Hi-fi, definition 21-2
*High pass filter 29
*IF 54
— — rejection 48, 54-5
— — signal boosting 54-5
— — strip and detector 54-5
*IHF 52
— — usable sensitivity 48, 51-2
*Impedance 34
*Inductance 140
— — symbol 136
Inductor 140
*Inertia (Pick-Up arm) 79-80
*Infinite baffle cabinet 122
Input—amplifier 32, 38-42
— — speakers 125
— — tape recorder 89, 106
— — tuner 48-52
Input plugs, fitting 152-4
Input/Overload ratios 41-2
Instant playback 91
Interference 46
Ionosphere 46-7
Jack plug—fitting 154
Jack socket symbol 136
Japan Victor Co. quad 86
*Jointing block 110
*Kilo 17-8
Kits 140-1
*Lacquer, cutting 62-3
Lifting lever 78, fig. 23
*Load—cartridge 82, 84
— — speakers 125-7
— — tape recorder 89-91
*Logic circuit 87
*Logic control 87
Long wave 46, 49
Loudness (Acoustics) 20-1
*Low pass filter 29
Mains transformer 140
— — symbol 136
Maintenance 157-70
— — cassettes 167-70, fig. 53
— — tape recorder 164-7, figs. 49-51
— — tapes 167-70
— — turntable 158-9
*Mass, effective (Pick-Up arm) 78-9
*Master disc, cutting 62-3
*Master shell 65
Master tape production 62
Matching 31-2, 35, 38-9, 51, 78, 84, 90-1, 95, 106, 116, 121, 125-7
*Matrix 85-6, fig. 28
Medium wave 46, 49
*Mega 17-8
*Micron 107
Microphone 104-6, fig. 34
*Milli 38
*Mixer, Mixing console 60-2, 100-1, fig. 18
*Modulation 44-5, 47, figs. 13-4
Monitoring 100
*Mother 65
*Moving coil—headphones 128
— — microphone 104-6, fig. 34
— — speakers 118
— — — — positioning 124
Multimeter 142-3, fig. 43
Multi-tracking 62
Music centre 22-3
Musical instruments—frequency ranges fig. 5
*Negative disc 65
Nippon Columbia Co. quad 86
*Ohm 32-3
Open-reel recorder *see* Tape recorder
*Oscillator 54
Output—amplifier 31, 33-5
— — cartridge 82, 84
— — speakers 125-7
— — tape recorder 89-91
— — tuner 48, 52-3
Overload capacity 40-1

*Overtones 18-9, figs. 5-6
Parallel tracking arm 76
*Phase response 120
*Phon 21
Phono plug fig. 46
—— fitting 154
Piano—frequency range 17, figs. 4-5
Piano-music, as a test 96, 117
Pick-Up 75-84
—— arm 75-80, figs. 21-3
—— choosing 75
—— cartridge 80-4, figs. 23-6
*Pitch (Acoustics) 14-8
Pivot (Pick-Up arm) 77, fig. 22
Pixall drum 160
*Platter 69-72
—— cleaning 158-9
Playback head 94-5, fig. 29
Pliers 142
Plugs—fitting 150-5, figs. 44-6
—— power 132, fig. 37
*Polar response 106
*Post-Echo 106-7
Potentiometer 136-7
—— symbols 136
Power-Amp 30
*Power handling (Speakers) 125-6
Power output—amplifier 31, 33-5
Power response 39
*Pre-Amp 30
—— tape deck 92
*Pre-Echo 106-7
Pressure roller (Tape recorder) 96, fig. 29
*Print through 107
Quadraphonic Stereo (QS) 85, 88
*Quadraphony 84-8, figs. 27-8
—— available systems 85-7
—— choosing a system 87-8
—— discrete systems 86-7
—— incompatibility of systems 88
Radio *see* Tuner
Radio transmission 45-7, fig. 2
—— interference 46-7
Radiogram 22-3
*Rarefaction 15, fig. 3
RCA quad 86
*Reactance 34
Record-changing systems 73-5
Recording—amateur *see* Tape Recording
—— commercial 58-65, figs. 17-9
—— —— disc production 62-5
—— —— tape 59-62
Recording head (Tape recorder) 94, fig. 29
Recording Industry Assn of America 39, 64
Records—cleaning 159-62, figs. 47-8
—— neutralising static 159
—— storage 157-8, 163
Rectifier 139
—— measuring 143
—— symbol 136
Reel capacity 89-90, 107
*Resistance 33, fig. 10
Resistor 136-7
—— measuring 142
—— symbols 136
Reverberation (Tape recording) 62, 100
*RIAA 39, 64
Ribbon microphone 105-6
*Root mean square (rms) 34, fig. 11
Rubber drive-wheel system 69
*Rumble 68, 73
Safety—electrical 132-4, fig. 37
Sansui quad 85
Screw-terminal speaker plug —fitting 152
Selecting an outfit 22-6
—— amplifier 23, 28, 31, 41-2
—— headphones 128-9
—— kits 140-1
—— pick-up 75
—— quad 87-8
—— speakers 25, 116-7
—— tape recorder 24, 91-2, 96-7, 113
—— tuner 23, 43-4, 47, 55
—— turntable 23-4, 58, 66-7, 73-5

*Sel-Sync 99–100
Sensitivity 32, 38, 42, 51–2, 125, 127–8
*Sensitivity field 106
Servo system (Turntable) 70–1, fig. 20
Shibata stylus 82–3
*Shim 93–4
Short wave 46, 49
*Signal to noise ratio 31, 37–8, 48, 53, 89, 94–5
Socket symbol 136
Solder 142
Soldering 148–55
—— iron 141
—— and plug-fitting 148–50
—— —— —— tools 149
Sound booster 119
*Sound on sound 98–9
Sound recording—amateur 98–101, 110–3
—— commercial 58 62, figs. 17–8
—— —— sound isolation 60
Sound reflection (Speakers) 123
Sound theory (Acoustics) 12–21, figs. 1–7
—— coloration 18–9, 117, figs. 5–6
—— intensity 20–1, fig. 7
Spacer 93–4
Speakers 25, 35, 115–28, fig. 36
—— cabinets 121–2
—— choosing 25, 116–7
—— cones 118–21
—— electrostatic 118–9
—— factors affecting sound quality 121, 123–5
—— importance 25, 115–6
—— input 125
—— matching 31, 35, 121, 125–7
—— moving coil 118
—— performance 127–8
—— plugs—fitting 150–2
—— positioning 122–5, fig. 36
—— specifications 125–8
—— —— example 125
—— types 118, 120
Specifications—amplifier 30–42
—— cartridge 81–4
—— pick-up arm 78–80
—— speakers 125–8
—— tape recorder 89–91
—— tuner 47–55
—— turntable 67–73
—— typical examples 31–2, 48, 68, 78–9, 81–2, 89, 125
Speeds—tape—cartridge 109
—— —— cassette 101
—— —— open reel 89–90, 96–7
—— —— —— choice of 97
—— turntable 57, 68
—— —— stabilisation 70–2, fig. 20
Splicing 110, fig. 35
*SQ 85, 88
*Squawker 119
*Stamper 65
Static electricity—records 159
*Stereo 85
—— Quadraphonic (SQ) 85, 88
—— radio reception 47, 53
—— separation 48, 53, 82, 84
—— speaker positioning 122–5, fig. 36
—— tape tracks 98
Stereo-record and playback 91
Storage, in general 157–70
—— cassettes and tapes 167–8
—— records 163
*Stylus 81–3, 162, figs. 23–6
—— checking 83, 162
—— configuration 81–3
—— diamond 83
—— sapphire 83
—— tracking force dial 78
*Superheterodyne 54–5, fig. 16
Switch symbols 136
Symbols (Circuit) 136
*Synchronous motor 70
Tape deck see Tape recorder
Tape heads—care of 165–7, figs. 49–51
—— cassette 102, fig. 32
—— open reel 89, 93–6, fig. 29

—— —— erase head 95
—— —— playback head 94–5
—— —— recording head 94
Tape monitor control 30
Tape record outlet 32, 39–40, 106
Tape recorder 61–2, 89–107, 113, 164–7, figs. 29–33
—— care of 164–7, figs. 49–51
—— choosing 24, 91–3, 96–7, 113
—— heads 89, 93–6, 102, 165–7, figs. 29, 32
—— input 89, 106
—— multi-track 61–2
—— objectives 93
—— output 89–91
—— specifications 89–91
—— —— example 89
—— speeds—cassette 101
—— —— open reel 89–90, 96–7
—— transporter 96–7
—— types 90–2
Tape recording 59–62, 98–101, 110–3
—— balancing tracks 99–100
—— commercial 59–62
—— direct 106–7
—— double tracking 98–9
—— echo 100
—— hints on recording 110–2
—— mixing sounds 100–1
—— monitoring 100
—— skills 110–2
Tape transporter 93, 96–7
—— weight balance 96
Tapes—care of 167–70, figs. 52–3
—— cartridge 109
—— cassette 108
—— chromium dioxide 108–9
—— cleaning 167
—— editing 109–10, fig. 35
—— manufacturers' recommendations 95
—— matching 95
—— open reel 107–8
—— preparing new tapes for use 168
—— splicing 110, fig. 35
—— storage 157–8, 167–8, fig. 52
—— tracks 98, figs. 30–1
Timbre 18–9, figs. 5–6
Tools 141–3, 149
Tracking—double 98–9
—— multi 62
Tracking error, maximum 79–80
*Tracking force 77
—— range 82–3
—— scale accuracy 79–80
Tracks 62, 98, figs. 30–1
—— balancing 99–100
—— specification 89
*Transcription deck 72
*Transducer (Tape recorder) *see* Heads
Transformer 140
—— symbol 136
Transistor 137–9, fig. 40
—— symbol 136
Treble control 29
Tuner 43–55, figs. 13–6
—— AM or FM? 47
—— choosing 23, 43–44, 47, 55
—— dial fig. 15
—— input 48–52
—— output 52–3
—— plugs—fitting 153
—— sensitivity 51–2
—— specifications 47–55
—— —— example 48
—— superheterodyne 54–5, fig. 16
Tuner-amplifier 43–4
Turntable 57–8, 66–75, 158–9
—— care of 158–9
—— choosing 23–4, 58, 66–7, 73–5
—— drive systems 68–70
—— importance 57–8
—— pick-up 75–84
—— platter 69–72
—— record-changing systems 73–5
—— speeds 57, 68
—— —— stability 70 1, fig. 20
—— specifications 67–73

—— —— example 68
*Tweeter 119
*UD-4 86, 88
Universal Discrete Four
Channel (UD-4) 86, 88
Variable capacitor 139
—— symbol 136
Variable resistor 136-7
Variable-speed drive 69-70
*Vari-Matrix 87
VHF 46-7, 49
Vibration (Acoustics) 12-21, figs. 1-7
Voice, frequency range 16
*Volts 33, 133, fig. 10
Voltage selector 134
*Watts 32-3, 133
—— DIN 125-6
—— rms 34, 125-6
*Wavelength—acoustics 14, fig. 2
—— radio 45-7
*Weighted measurements 37-8
Wire strippers 142
*Woofer 119
*Wow 68, 72, 89-90, 96
Zerostat gun 159-60